ELECTROMAGNETIC POLLUTION SOLUTIONS

What You Can Do To Keep Your Home & Workplace Safe

Dr. Glen Swartwout

AERAI Publishing

Hilo, Hawaii

ELECTROMAGNETIC POLLUTION SOLUTIONS

≺ TABLE OF CONTENTS ≻

Dedication

**THIS BOOK IS DEDICATED TO
SEVEN GENERATIONS
AS YET UNBORN.**

In the traditions of the native peoples
of North America, the responsibility
rests with us to deliberate and act
on behalf of these next seven
generations.

By understanding the great implications
of our actions today, we can contribute to a future where technology and
humanity progress with absolute
synergy.

When we grasp the full scope of the
electromagnetic paradigm in health,
development and performance, we
hold the key to that future.

*This guide is intended
to bridge the gap
between the wisdom
of human experience
and the highly technical
information of pure science
which generally ignores
the difficult issue
of human impact.*

➤ A young boy had been diagnosed with leukemia. He was receiving the best care of both the orthodox and the natural medical practitioners. His naturopathic doctor, one of the author's mentors, found through Bio-Energetic Regulatory tests that electromagnetic stress was a causative factor in this case. The family lived well within a quarter mile of one of the large transmission lines that course through Toronto. Being skeptical about the significance of the electromagnetic stress, the family was taking a wait and see approach, while doing everything else possible to help the boy. During this time, the leukemia continued to worsen, although it was a slow decline because of the good medical and naturopathic care. Then, during the summer, the family unknowingly conducted a clinical trial of moving the boy out of the electromagnetic stress field. He had gone to stay with his grandmother for a couple weeks. For the first time in his treatment, he showed improvement in his blood count. When this improvement began to fade rapidly, the family was convinced of the value of further investigation. Again the boy was sent off to grandma's, and for the second time showed improved response to his treatments. Finally, this family had gained enough personal experience through their own investigations to confirm the original Vegatest findings of their N.D. In this case, it was actually necessary for the family to sell this house and move to one in a less polluted electromagnetic environment. This was a small price to pay for the health of their son, however.

❖

In physics, as first proposed by Einstein, there is ultimately one unified energy field. This field has been perceived and comprehended in this century in the forms of the four forces of physics. These are the strong and weak forces of the atomic nucleus, the force of gravity which acts on the scale of the heavenly bodies, and electromagnetism, which governs interactions on the molecular scale. The process of life depends entirely upon electromagnetic interactions between the molecules which make up our bodies. The making and breaking of chemical bonds is electromagnetic. The release of energy from food is electromagnetic. When two electromagnetic fields are superimposed, the interaction is called interference.

What is Electromagnetic Stress?

Electromagnetic stress is metabolic disruption caused by electromagnetic interference. This can be due to artificial or natural electromagnetic fields. As these fields interpenetrate the tissues of the body they alter the electromagnetic functions of the body. Because electromagnetic fields are generally silent, invisible, odorless, and tasteless, they usually go unidentified as a source of stress.

Electromagnetic stress is metabolic disruption caused by electromagnetic interference.

TABLE 1.

Stress Related Diseases [1]

Accidents

Addiction

(e.g. nicotine, caffeine, sugar)

Alcoholism

Allergies

Anti-social behavior

Anxiety

Arthritis

Asthma

Back pain

Cancer

Chronic viral syndromes

Constipation

Depression

Diabetes

Drug abuse

Eating disorders

(Anorexia, Bulemia, etc.)

Frequent infections

Glaucoma

Headache

Heart Attack

Hypertension

Hypoglycemia

Indigestion

Most chronic conditions

Muscle cramps and tension

Obesity

Psychoses

TMJ problems

Ulcers

Vision problems

Electromagnetic stress is common in today's world of computers, telephones, electric lights, and the array of electrical gadgetry. Because of the increased use of electricity in the past century, electromagnetic interference even shows up high in the ionosphere. Anomalies in the ionosphere were traced to the electrical grid here on earth because they show up on the weekends when more power remains in the grid to produce resonance effects at that distance. These anomalies, like the grids that induce them, oscillate at 60 Hz over the U.S.A. and 50 Hz over Europe. Electromagnetic stress affects our energy and vitality, leaving us unable to cope with the normal processes of life, including detoxification, rejuvenation and regeneration, which are normal daily bodily functions especially during sleep. The reason we are often affected at night is simple: we leave our body in one location in space for many hours, exposing it to whatever the ambient electromagnetic fields may be in our bedroom, including those put out by whatever is plugged into the wall sockets. At the same time, the body at rest produces a smaller, more sensitive field of its own.

Of course there are occupations that can unknowingly expose us to harmful fields as well, not to mention entire schools and homes which are exposed to high levels of electromagnetic radiation from power lines. To understand the scope of the problem and the range of solutions requires an understanding

Electromagnetic stress is common in today's world of computers, telephones, electric lights, and the array of electrical gadgetry.

Our occupations, schools, and homes can unknowingly expose us to high levels of harmful electromagnetic radiation.

The human body receives electro-magnetic signals from its natural environment as well as from itself.

of how fundamental the electromagnetic spectrum is. Table 2 illustrates the range and types of Electromagnetic Field (EMF) oscillations to which we are exposed, both from natural and man-made sources.

Picture a piano keyboard with 6 octaves. Now imagine a keyboard 10 times as long. That long keyboard represents the part of the electromagnetic spectrum we know. Each note is like a radio or television station, broadcasting from its own spot on the dial. Like the familiar electromagnetic broadcasts of radio and tv, each channel has its unique programming, with that musical note carrying with it a particular stream of information. When we pick up two channels at the same spot on the dial, there is interference. It is impossible to get meaning out of the information broadcast to us.

Electro-magnetic stress can interfere extensively with our natural program-ming.

The human body receives electromagnetic signals from its natural environment as well as from itself. These range from energy transfers and biochemical reactions on the molecular level, to sensing the changing seasons by the length of daylight from the sun. The light we see from the sun is all contained within one octave of that full keyboard. Other biologically important notes extend out over at least 32 octaves. It is mind boggling to realize just how extensively electromagnetic stress can interfere with our natural programming.

TABLE 2.

Electromagnetic Frequency Spectrum

Nature's frequency ranges:	And some new uses:
• 10^{23} Hz atomic nuclei, cosmic radiation	• radioactivity, fallout
• 10^{20} Hz gamma rays	• medical and military gamma rays
• 10^{19} Hz x-rays, atomic resonances	• medical x-rays
• 10^{18} Hz x-rays, atomic resonances	• medical x-rays
• 10^{17} Hz ionizing ultraviolet, molecules	• increasing due to ozone depletion
• 10^{16} Hz non-ionizing ultraviolet, macromolecules (protein, enzymes, viruses, DNA)	• deficient in artificial lighting, absorbed by window glass and eyeglass lenses
• 10^{15} Hz near UV, bacteria	• deficient in artificial lighting
• 10^{14} Hz visible light spectrum, temperature of the sun, cellular resonances	• distorted in artificial lighting, with deficiencies and excesses in specific ranges
• 10^{13} Hz infra-red (heat), fungi, water	• EHF: millimeter wave communications, heat sources
• 10^{10} Hz molecular resonances	• SHF: microwave radio, ovens
• 10^{9} Hz individual organ resonances	• UHF: airport, police, military radar
• 10^{8} Hz total human body resonance	• VHF: FM, TV broadcasting
• 10^{7} Hz	• HF: short wave radio
• 10^{6} Hz	• MF: AM broadcasting
• 10^{5} Hz	• LF: navigation beacons
• 10^{4} Hz thunderstorms, high pitch sound	• radio telegraphy, noise, CRTs
• 2×10^{3} Hz nuclear magnetic (proton) resonance (NMR) with geomagnetic field (GMF)	• VLF, noise
• 10^{3} Hz muscle & nerve potentials	• VLF, noise
• 5 or 6×10^{1} Hz low pitch sound	• electric power systems
• 10^{1} Hz brain waves (EEG), earth's Schumann field	• ELF; vibration; VDTs; television
• 10^{0} Hz heart beat (EKG)	• ULF
• 10^{-1} Hz physiological functions, geophysical oscillation	• ULF
• 10^{-4} Hz	• ULF
• 10^{-5} Hz day/night cycle of earth's electromagnetic field	• ULF
• 0 Hz	• static fields (e.g. magnets and capacitors in television sets, stereo speakers, telephone receivers, etc.)

Some information is included to help guide the more serious investigator or thinking clinician and may be more detailed than you require.

Some information is included to help guide the more serious investigator or thinking clinician and may be more detailed than you require in order to solve your immediate problems. Please skim over anything you find too technical at this time, but keep this guide in mind for future reference by yourself or your holistic health care provider. Since knowledge and solutions are ever developing, if this is an area of ongoing concern to you, or if you are interested in other factors which impact on human health, performance and development, then membership in our Research Academy may be of interest to you as well.

FOOTNOTES

[1] Andrew, R.J., **Beyond Disease, The Energy Economics of Preventive Health**, (Reno, NV: Earthly Promotions, 1983).

➤ A physician in Oregon who himself has diabetes was in the habit of wearing a beeper on his left side. Neither he nor his personal physicians realized the potential significance of this source of electromagnetic fields so close to the body. When he developed a chronic inflammation in his left foot, however, nothing would get rid of it. He tried many treatments, both natural and allopathic, yet still the inflammation continued. Only when the author identified the connection between the location of the beeper and its electromagnetic stress field on precisely the same meridian as the inflamed area of the foot was improvement possible. The acupuncture meridians carry not only a flow of energy, but also metabolic waste products. The flow is driven by the body's own electromagnetic field. When a much stronger field is superimposed, it can block the flow of the meridian, resulting in energy imbalances, functional disturbances and toxicity buildup. Over longer periods of time, clinical disease is the result.

➤ In 1987, the author's wife, K. Aubrey Swartwout, began noticing sleep disturbance, especially difficulty falling asleep without any apparent reason. This occurred after putting a touch-on lamp on the night stand. The problem was identified through a Vega test electro-magnetic scan during routine research, and removal of the lamp from the bedside resulted in immediate relief.

❖

This section will present the most common sources of both natural and man-made electromagnetic interference. The emphasis is placed on those sources of stress most likely to be affecting you.

Types of Electromagnetic Stresses

Hazard Ratings given in this section are based upon clinical experience with hundreds of patients with electromagnetic stress problems. They are intended to provide a relative guide to the probable significance to you of each source. Each rating is based on such factors as field strength, length of exposure, and number of people affected. The range is from 10 (greatest hazard) to 1 (least likely to be problematic.) These ratings follow the source of interference, as in the following example: **Local distribution lines and high tension transmission lines** *(10).*

MAN-MADE STRESSES

Most man-made electromagnetic fields are either 60 Hz fields as in the USA or 50 Hz fields as in Europe and the USSR. These are the frequencies of oscillation of the electrical utilities' power grids, to which most people are connected.

Most man-made electromagnetic fields are generated by electrical utility companies.

Electrical utility distribution system:

Studies are beginning to show the increased risks of conditions as serious as leukemia and cancer through exposure to electromagnetic radiation from electric utility sources.

Local distribution lines and high tension transmission lines *(10):* Local electrical distribution lines can radiate fields about 150 feet in all directions. The larger power lines radiate significant electromagnetic fields up to a half mile wide (a quarter mile to each side). Studies are beginning to show the increased risks of conditions as serious as leukemia and cancer with this type of EMF exposure.

Substations *(10):* These substantial installations radiate stress fields for a quarter mile in every direction. It is very wise to consider their location as well as that of any high tension giants discussed above when making real estate decisions. It is likely that, as this information becomes more widely known and accepted, real estate values will begin to reflect these problems that so easily traverse property boundaries. Don't invest either your money or your health on a bad risk.

Transformers *(8):* Smaller end-of-line transformers, like the one that is likely in front of the building you are in, radiate fields of 150 feet or more, varying with the different types in use.

Fuse boxes *(8):* When located in the bedroom or workspace or even when situated inconspicuously on the other side of a wall, they are a source of stress. Fuse boxes almost always have some electricity flowing through them, so they are a source of very chronic stress when implicated.

Construction methods:

Concrete foundations *(10):* block the normal and beneficial Schumann field. Concrete is also a source of ionizing radiation. It is therefore important, since most of us both live and work in such an environment, to get outdoors daily in a natural area, such as walking on a lawn or at the beach.

Metal building materials *(9):* These also block the beneficial Schumann field and can act to amplify harmful EMF.

Old or poorly designed wiring *(8):* Old styles of wiring, such as aluminum wiring, not only create fire hazards, they also increase electromagnetic stress. One old type of wiring runs bare wires on adjacent studs, using porcelain insulators. Because of the separation of the two currents running in opposite directions, this design greatly increases magnetic fields. The lowest magnetic fields are produced by heavy guage wires that are twisted tightly around each other. The low resistance and the twisting both help to increase efficiency and reduce magnetic fields. The proximity of wiring in the walls to sleep and work spaces is also important, since fields drop off with distance. Wiring design is especially important in the bedroom, because of

> It is important to get outdoors daily in a natural area, such as walking on a lawn or at the beach.

> Old styles of wiring, such as aluminum wiring, create fire hazards and increase electromagnetic stress.

The proximity of wiring in the walls to sleep and work spaces is important.

the sensitivity of the body to electromagnetic fields during sleep. Wires to bedroom wall plugs can be run down from the ceiling rather than along the wall at plug level. Also, a separate circuit breaker switch for the bedroom circuits may be installed in the bedroom. In this way, even the residual fields resulting from the resonance of alternating current in the wiring system can be eliminated during sleep. This can be especially important if the bedroom circuit wiring surrounds the bedroom on 3 or 4 sides. Step by step instructions for correcting or designing a good electrical system are available on the video **Current Switch.**[1]

Electrical systems grounded to water pipes *(7):* This problem extends to neighboring homes on the same water line, when there is a short circuit or other faulty wiring. Beyond the potential for chronic electromagnetic stress, improper grounding also increases the risk of electrocution during lightning storms. This is a form of acute electromagnetic stress.

Make sure all wiring is properly grounded and that the polarity is correct.

Faulty wiring *(7):* When the polarity of the wires are reversed in household or office wiring, extra energy is lost to production of electromagnetic fields. This means increased health risks, as well as higher utility bills. Reversed polarity current, caused by improper connection of the wires to a wall plug, also puts increased wear and tear on electrical appliances.

Electrical appliances in the household:

The typical situation of man-made electromagnetic stress occurs in the average modern bedroom. During sleep, the body's functions go into a resting and recuperating phase of activity. The body's own electromagnetic functions become more sensitive to fields. This makes sense when you realize that the sun is the source of most natural EMF on earth. At night we are on the side of the earth that is shielded from that exposure. Our bodies, therefore, are not designed to overcome EMF interference at night.

During sleep, the body's own electromagnetic functions become more sensitive to external electromagnetic fields.

The functions most interfered with are those most active at night. These include certain channels in the body called meridians, which have been proven to carry direct electrical current. The meridians active at night are those regulating the liver, gall bladder, lungs and large intestine. The normal detoxification process is blocked by the effects on all of these meridians and their associated organs.

Mental integration, rest, and recuperation are also blocked by the false state of 'sleep' which these electromagnetic fields induce. The normal dreaming cycle is disrupted or completely bypassed. EEG measurements of continuous low frequency brain waves indicate that it is more like being in an induced coma than like normal sleep.

External electromagnetic fields can disrupt or cause the body to completely bypass the normal dreaming cycle.

Due to both the mental and physical effects, such a disturbed 'sleep' pattern can become very addictive. The need for true sleep and rejuvenation increases with time.

Studies are showing significant health risks associated with the use of heated water beds, electric blankets and heating pads.

Heated water beds (10): most water beds contain heating coils to heat the water. When sleeping on such a bed the water transmits the field produced by the heating elements directly to the unwary sleeper. Normal sleep patterns are interfered with. If you must sleep on one, heat it during the day and unplug it at night. The water will hold the heat all night due to its high specific heat capacity. If you must buy a water bed, get one without a heating coil. They are currently available.[2]

Electric blankets and heating pads (10): The current in these devices is much too close to the human body. Studies are showing significant health risks with the use of these devices. The effects totally interrupt the normal mechanisms of the body's temperature and circulation regulation. This is obviously not helpful to anyone who is already suffering from poor circulation.

Cigarette smoke contains significant amounts of radioactivity.

Cigarette smoke (10): Cigarette smoke contains significant amounts of radioactivity because of the Uranium contained in fertilizers used in commercial tobacco growing. The Uranium decays to Radium and then to Polonium 210, which is taken up by the plant. As a result, the lungs of a pack-and-a-half-a-day smoker are exposed to the equivalent of 300 chest x-rays a year. Those who live with smokers are not only exposed to the radioactive Polonium, but also to increased levels of radioactive Radon breakdown products which are attracted to smoke particles. This causes Radon products to stay in the air rather than settling onto surfaces as they usually do. Cigarette smoke contains other radioactive elements as well: Radium 226, Lead 210, Bismuth 210 and Potassium 40.[3]

Microwave ovens *(9):* Microwaves break nitrogen bonds in proteins, so they are not recognizable to our digestive enzymes. Since most Americans eat 2 to 3 times too much protein they don't notice much difference in digestion, though. Even 'safe' levels of microwaves have unknown health implications. If you already own one and you insist on making some use of your monetary investment, the least problematic use is to preheat microwave safe utensils and water. This is what microwave ovens were originally intended for. Since units can begin to leak as they age, it is important to check them about twice a year using a detector, or better yet, a measuring device like the Trifield meter. Even in homes free of microwave sources, some resonance may be carried into the home through utility wires from homes on a common end-of-line transformer. This is another advantage to off-grid systems such as photovoltaic (solar electricity).

Micro-waves break nitrogen bonds in proteins, so they are not recognizable to our digestive enzymes.

Effects of radioactive ionization-type smoke detectors can extend up to 50 feet.

The head contains our most specialized EMF receptors: the retina and the pineal.

A color television set, even unplugged, maintains a charge for at least 3 days.

Smoke alarms *(8):* Effects of radioactive ionization-type smoke detectors can extend up to 50 feet. They contain Americium 241.[4] Do you have difficulty getting a good night's sleep in a hotel room? This is one reason: they are all equipped with a smoke alarm in the bedroom! Radiation-free smoke detectors, using a photoelectric process rather than ionization, are available and should be used.[5] Not only are they radiation-free, they even work better, according to Consumers Report. But please note, if you get rid of your ionization type smoke detector, please follow the disposal instructions on the unit. It contains radioactive material, and should not be thrown in the trash! (So why, then, would we want them in our homes?) Also remember that using a battery operated model minimizes the non-ionizing radiation given off, even by the photoelectric units which contain no radioactive substances.

Clock radios *(8):* Keep the cord and the radio at least 8 feet from your body, and especially your head. The head contains our most specialized EMF receptors: the retina and the pineal.

Color television *(6):* A color television set, even unplugged, maintains a charge for at least 3 days. Don't sit closer than 6 to 10 feet when watching, and limit your viewing time to less than 2 hours a day. The flicker of the light from the screen is in the alpha brain wave range. This is one of the main reasons that the "boob tube" is so named.

By looking at a television picture we are automatically entrained, or electromagnetically forced, into a relaxed, meditative, hypnotic, suggestible state. Also, old color TVs may emit x-rays.

Heating and cooling appliances *(5):*
Because items like electric heaters, stoves, toasters, refrigerators and air conditioners draw large amounts of current, they produce strong electromagnetic fields. Like all energy fields, they drop off with distance, so this only becomes a significant problem if we spend alot of time within a few feet of one of these appliances. At times this becomes a little tricky to realize, because magnetic fields go right through walls. Fields are also emitted by the wiring carrying the current to the appliance. It is especially important to check the location of appliances and wiring in walls that are shared by a bedroom and another room like the kitchen, where a refrigerator will be going on and off during sleep hours. The best way to rule out fields from unseen sources like wiring in walls, ceilings or floors is to monitor periodically with a simple ELF magnetic field meter[6].

Certain other building materials *(4):*
Phospho-gypsum based materials like anhydride are extremely high in radioactive Uranium (13.7 ppm) and Thorium (16.1 ppm). This material should definitely be avoided. Construction with granite, and sometimes brick or concrete can increase indoor radon levels by 50% over wood

Magnetic fields go right through walls. The best way to rule out fields from unseen sources like wiring in walls, ceilings or floors is to monitor periodically with a simple ELF magnetic field meter.

Wood is the best of all building materials in relation to harmful electromagnetic energy emissions.

Entire sections of the United States and other commercially farmed areas have had their topsoil polluted with radioactive ingredients found in chemical fertilizers.

Uranium found in tableware can be leached out, transferring the radiation source to the body.

construction. Granite averages 4.7 ppm of Uranium and 2.0 ppm of Thorium, while cement averages 3.4 ppm Uranium and 5.1 ppm Thorium.[7]

Commercial fertilizers *(4):* Phosphate fertilizers used extensively in agribusiness contain Uranium and its radioactive breakdown products. Entire sections of the United States and other commercially farmed areas have had their topsoil polluted with this ionizing radiation source.

Touch-on lamps *(4):* While these are handy on the nightstand, this is just the worst place for them. In order to function, they carry a constant low level of current through the lamp.

Ceramics *(3):* Some ceramics, including tableware, contain up to 20% Uranium compounds. These can be leached out, especially by acid foods like tomato or citrus, transferring the radiation source to the body. Be especially suspicious of shiny surfaces with colors like red, orange, yellow or beige.[8]

Shavers *(3):* While those who use an electric shaver are only exposed for a short time each day or periodically, the electromagnetic field levels to which the user is exposed are very high. Since alternatives are widely available, it is wise to use a rechargeable battery operated shaver, a wind up mechanical shaver, or the old fashioned razor blades.

Irradiated foods *(2):* While foods being treated with radiation do not retain the ionizing radiation they have been exposed to, there is much reason for concern over the effects of eating these 'dead' foods:

The energetic resonance of the radiation source stays in the food.

The energy of the radiation which is absorbed by the food stays in the food, dissipating over a period of weeks. If consumed too soon after irradiation, deleterious health effects are maximized.

The abnormal chemical breakdown products caused by ionizing radiation continue to be present in the food and are likely to include toxins and carcinogens.

Enzymes and nutrients in the food are destroyed, decreasing food value and promoting mal-digestion.

The life energy of the food is released in the form of biophotons, leaving a lifeless effigy of real food.

Normal bacterial flora are destroyed.

Pathogenic bacteria and fungi, like those producing botulism and aflatoxin survive. In addition, they are provided with a growth medium free of competition. Because the toxins produced by the pathogens have no smell, it is not possible for you to tell if the irradiated food has gone bad. The warning smells of putrefaction and fermentation are gone with the bacteria that would produce them.

While foods being treated with radiation do not retain the ionizing radiation they have been exposed to, there is much reason for concern over the effects of eating these 'dead' foods.

Even government inspection of irradiated foods as they enter the country cannot guarrantee their safety. They may contain acceptable levels of biological toxins at the time they are imported. Then, in storage, the pathogenic flora can thrive. Aflatoxin, one of the substances produced, especially in peanuts, is one of the most carcinogenic substances known. Another, clostridium botulinum, can cause rapid death from food poisoning.[9]

Personal, portable electrical devices:

Beepers, pagers *(9):* As personal communication devices like beepers and pagers become more common, the hazard from these increases in incidence. Beepers or pagers are especially injurious because they are often worn on the belt and are therefore in close proximity to the body, and certain meridians in particular. This most often involves one or more of the meridians that cross the waist line.

Cellular phones *(8):* The main source of stress with cellular phones is the transmitter, which must be powerful enough to transmit outgoing signals over the fairly long distances of the cellular network. This transmitter may be in close proximity to the user for several hours a day or more. This is compounded by the usual electromagnetic stress common to all telephones. That is due to both the electromagnetic speaker

Beepers or pagers are especially injurious because they are often worn on the belt and are therefore in close proximity to the body.

and microphone in close contact with the ear and mouth area respectively.

Watches *(6):* The battery driven electromagnetic oscillation of the piezoelectric quartz crystal in most watches today presents a potential interference with the many meridians which run through the wrist. Also, for those who wear a quartz watch pendant around the neck, the thymus, a most important element in the immune system, may be affected. Old "glow in the dark" or luminescent dial watches may contain Radium. Many of the people who painted those dials got mouth cancer from the practice of licking their brushes to maintain a fine point.[10]

Watches of various types each present an associated specific hazard.

Metal and synthetic materials in contact with the body:

Buccal currents *(9):* "Buccal" literally means the cheeks. The buccal currents are measurable electrical currents in the mouth caused by the effects of foreign metals in the body. These dissimilar metals, such as the silver and mercury combined in the common amalgam fillings, set up a battery effect. The metals are leached out into the tissues, causing many different types of health and performance problems.[11,12] Heavy metal ions are known to plate out on cell membrane.[13] They also substitute for other minerals with similar electron configurations (see Volume 2). Thus they block

Common silver and mercury tooth fillings can set up a battery effect in the mouth, or be leached into surrounding tissue.

important enzymes for which the nutrient minerals act as cofactors.

Pins, screws, and plates *(8):* Currents similar to the buccal currents can be set up in any part of the body by the presence of metal parts.

Synthetic clothing *(8):* Static electrical charges of up to several thousand volts are generated by normal movement when wearing synthetic clothing. The currents generated are enough to interfere with normal functioning of the body's sensitive electrical system.

Necklaces *(6):* Metal necklaces block the normal transmission of the electrical heart rhythm (EKG) to the head.[14] They can receive resonant frequencies of environmental electrical pollution and transmit them to the body, especially the thymus, which is an important part of the immune system. Necklaces can also interfere with the normal electromagnetic functions of the meridians which pass through the neck.

Jewelry *(4):* Each piece of jewelry picks up specific resonant frequencies from the environment, like an antenna, transmitting them directly to the body through the skin. This transmission to the body is especially easy if the jewelry has metal in contact with any of the hundreds of highly conductive acupuncture points located all over the body. The acupoints are especially concentrated on the ears, the fingers and the toes. So rings and earings almost always

Static electrical charges of up to several thousand volts are generated by normal movement when wearing synthetic clothing.

Necklaces and other jewelry can attract harmful electromagnetic radiation to the body.

affect at least one acupoint. In fact, earings were first worn by sailors to stimulate an acupuncture point on the ear that affects the eyes. Gold has a tonifying or stimulating effect, so this metal was worn to stimulate improved eyesight, so important for navigation at sea. Of course many earings are made of alloys containing potentially toxic metals, so the effects on vision today are "less clear."

Metal eyeglass frames *(4)*: The metal connection in these frames short circuits the brain's hemispheres where a difference exists in the electrical potential of the two sides of the brain. This can contribute to mental confusion and fatigue as well as headaches for those with a strongly dominant hemisphere.

Glass eyeglass lenses *(3)*: Optical glass may contain Uranium and Thorium, exposing the eye to ionizing radiation that can contribute to cataracts.

Eyeglass prescription *(6)*: Lens prescriptions affect the intensity and distribution of electromagnetic energy on the retina. The retina is specifically designed to receive light which is one octave of the electromagnetic spectrum. The retina is the primary sensory receptor of the body. Each retina has the capacity to send as much information and electromagnetic energy to the brain as all other sensory systems combined. This electromagnetic nerve current is transmitted in the form of waves of

Earings were first worn by sailors to stimulate an acupuncture point on the ear that affects the eyes.

Eyeglasses have numerous effects on the body in relation to electromagnetic energy.

electrical action potentials. Nerve current is what stimulates and regulates the brain and all of its functions. The lens prescription which minimizes electromagnetic stress on the retina and brain may not give the sharpest vision. Vision that is too sharp reduces the total information flow and increases stress.[15]

Ozone depletion *(6):* UV radiation reaching the surface of the earth is increasing. This is caused by chlorinated fluorocarbons emitted by refrigeration plants, leaking refrigeration systems, and many spray cans. This means that in the future, excessive exposure to sunlight will be even more harmful than now. The incidence of UV related diseases like skin cancer and cataracts will continue to rise. It will become much more important to maintain a truly healthy and clean body, eliminating those toxic and synthetic chemicals from the body which become even more detrimental when energized by the absorption of radiant light energy from the sun.

Military, medical and industrial waste radiation *(4):* Every state has some kind of military or industrial nuclear facility. In fact, 90% of the American population lives within 100 miles of a nuclear power plant. About one and a half million Americans are exposed to ionizing radiation at their workplace. Health effects have been shown to begin at levels at least as low as *30 times below* current regulatory standards. As soon

The incidence of UV related diseases like skin cancer and cataracts will continue to rise.

90% of the American population lives within 100 miles of a nuclear power plant.

as these findings were discovered, however, the Department of Energy stopped funding the research and failed to change allowable exposure limits.[16] Hospitals also produce significant radioactive waste. Dump sites containing these materials present an ongoing hazard, since the half-life of some radioactive materials is many thousands of years. High exposures also result from being downwind of testing and accidents. Wastes of course are transported along major highways and rail lines, so accidents here pose an additional potential for exposure of millions of people.

NATURAL STRESS SOURCES

Geopathic stress is a general term for naturally ocurring electromagnetic stress. The origin of the term is from earth (geo) fields, which over prolonged periods of exposure, produce disease (pathic). European investigators have demonstrated relationships between these fields and multiple sclerosis, rheumatism, arthritis, cardiovascular dysfunction, and cancer. Dr. Hans Nieper, an oncologist and Director of the Silbersee Hospital in Germany, has shown that at least 92% of his cancer patients have experienced chronic exposure to geopathic stress, especially in their sleeping place.[17]

European investigators have demonstrated relationships between naturally occuring electro-magnetic fields and multiple sclerosis, rheumatism, arthritis, cardiovascular dysfunction, and cancer.

The Schumann Field, Hartmann Grid and Curry Grid are examples of fields produced by the Earth which have effects on the human body.

Schumann Field: The Schumann Field is a beneficial field, which by its deficiency causes much stress, dysfunction and symptom-atology. Early in the space program, NASA discovered that many of the symptoms of space sickness experienced by the astronauts were due to the absence of these important Schumann resonances. Consequently, since that time, devices have been installed in all manned space flights that simulate the Schumann field oscillations, thus eliminating much of the stress of space flight. Here on earth, however, many people have very little exposure to these important electromagnetic frequencies, because blacktop, concrete, steel, and lowered water tables interfere with the normal resonance between the ionosphere and the conductive earth.

Hartmann Grid: The Hartman Grid is a repetitive resonance or standing wave pattern over the surface of the earth. The energy of a single grid line is usually not clinically signifi-cant. The lines of the grid are oriented in two orthogonal directions.

The lines, called HM2, are spaced every 2 meters in one direction and every 2.5 meters in the other direction. In both direc-tions, they have a width of 20 centimeters.

Every 10 meters in both directions there is a stronger, pathogenic grid line. This is called the HM3, and it is 40 centimeters wide.

There are also lines called HM7 and HM11 which are somewhat pathogenic. HM7 is 1.2 meters wide and HM11 is 2 meters in width.

Curry Grid: The Curry Grid is another resonance pattern of standing waves which covers the globe. It is oriented at 45 degrees to the Hartmann Grid and its lines also travel in two orthogonal directions. The lines of the Curry Grid are spaced 7 meters apart. Where its crossing points intersect those of the Hartmann grid, double grid zones of very strongly pathogenic fields occur.

Underground water channels

Announcement lines: These are generally benign, small fluctuations in the earth's electromagnetic field. They are equally spaced to either side of the stronger field found directly above the moving underground water channel.

Head lines: The zone directly, vertically above the moving underground water stream has a stronger field, and is the most pathogenic.

Center lines: The central zone above the part of the stream where the water speed is greatest produces a moderately pathogenic field.

Underground water channels and faults in the Earth generate small fluctuations in the Earth's electromagnetic field.

Fault zones

Announcement lines: These are generally benign, small fluctuations in the earth's electromagnetic field. They are usually asymmetrically spaced on either side of the fault. Spacing depends upon the angle at which the fault dips into the earth.

Fault line: The zone where the fault meets the surface of the earth carries the greatest and most detrimental electromagnetic field variations.

Geopathic disturbance fields are either charging (yang) or discharging (yin). **Charging fields** tend to produce sympathetic nervous system responses such as hypertension, stroke, heart attack, insomnia, hyperactivity, mania, alcoholism, migraine, childhood epileptiform fits, and nervousness. **Discharging fields** tend to produce responses of the parasympathetic nervous system including low energy, fatigue, arthritis, multiple sclerosis, malignancy and other degenerative disorders.[18] There is also a tendency towards reversal of the spin characteristics of protein molecules.[19]

Geopathic disturbance fields are either charging (yang) or discharging (yin). Each produces different sympathetic responses in the human body.

Sunlight: Excessive amounts of sunlight can be received by those who spend many hours outdoors, especially near snow or water, in the tropics, or at high elevation. This not only causes sunburn, but also predisposes one to skin cancer and other tissue degeneration. The dark tans that some people prize block the normal ability of the skin to produce vitamin D, essential to absorption of calcium. *Much of the damage done is due to the presence of toxic or potentially toxic (phototoxic) synthetic or pharmaceutical compounds not belonging in the body.*

UV radiation: The Ultraviolet (UV) part of the solar spectrum includes both ionizing and non-ionizing radiation. Ionizing radiation splits up molecules directly when they absorb its photons, or energy packets. Non-ionizing radiation, in contrast, can only do harm either by building up heat faster than the circulation can carry it away, or by energizing toxic substances present in the body. The retina in the eye can be most easily burned by heat build-up, when looking directly at the sun, because the macula or center part of the retina has no capillary circulation to efficiently remove excess heat. Synthetic chemicals that are not a normal part of the human physiology become more toxic when they absorb light. Sunbathers are 700% more likely to develop skin cancer by eating margarine versus butter.

Radon gas : Radon gas (e.g. in shale bedrock) is a common source of household and workplace radioactivity, especially in newer, tighter buildings. In areas where high levels of radon gas escape the bedrock, as in many permeable shales, the tradeoff of radioactivity

> The Ultraviolet (UV) part of the solar spectrum includes both ionizing and non-ionizing radiation, both of which can contribute to harmful effects on the body in excess.

> Radon gas (e.g. in shale bedrock) is a common source of household and workplace radioactivity

The Earth's electro-magnetic fields are affected by a variety of naturally-occurring phenomena.

for energy efficiency may not be worth the health costs.

Geologically young granite bedrock (e.g. Conway Granite in NH) shows a higher background radiation level and is one factor in the higher cancer rates in such areas.

Ore deposits cause local anomalies in the intensity and direction of the earth's magnetic field. Ferromagnetic ores take on the same polarity as the magnetic field they are in, thus increasing the earth's field in their vicinity. Paramagnetic ores take on an opposite field to that of the earth, reducing the resultant field. Diamagnetic rocks have no response to magnetic fields.

Charge variations in the upper atmosphere cause variable shifts in the earth's natural fields. These are affected by:

Ozone depletion.

Solar polarity variations in the sun's interplanetary magnetic field (IMF), which the earth passes through in its orbit.

Solar magnetic storms, which wax and wane in number on an eleven year cycle.

Moon Phases: Changes occur due to the gravitational variations associated with the moon's phases.

Magnetic and gravitational field variations with location on the earth due to:

Latitude, with weaker magnetic fields toward the poles.

Altitude, with higher gravity at low elevation.

FOOTNOTES

[1] **Current Switch,** (available for $50 from AERAI at (800) 788-2442).

[2] e.g. Comfort Sleep Serenity™ by Polcor (contact your local waterbed dealer for further information).

[3] Shannon, S., **Diet for the Atomic Age, How to Protect Yourself From Low-Level Radiation,** (Wayne, NJ: Avery Publishing Group, Inc., 1987).

[4] ibid.

[5] $25 for battery operated unit from Baubiologie Hardware at (800) 441-8971.

[6] The simplest low cost detector is the Power Pet, which costs $50 from AERAI at (800) 788-2442.

[7] Poch, D., **Radiation Alert,** cited in Chaitow, L., and Kutter, E., **How to Live With Low-Level Radiation, A Nutritional Protection Plan,** (Rochester, VT: Healing Arts Press, 1988).

[8] ibid.

[9] Webb, T., Lang, T., and Tucker, K., **Food Irradiation, Who Wants It?.** (Rochester, VT: Thorson's Publishers, Inc., 1987).

[10] Poch, D., **Radiation Alert,** cited in Chaitow, L., a **to Live With Low-Level Radiation, A Nutritional Pr** (Rochester, VT: Healing Arts Press, 1988).

[11] Stortebecker, P., **Dental Carries as a Cause of Nervous Disorders,** (Orlando, FL: Bio-Probe, Inc., 1986).

[12] Stortebecker, P., **Mercury Poisoning from Dental Amalgam - a Hazard to Human Brain,** (Orlando, FL: Bio-Probe, Inc., 1986).

[13] Habib, M.A., and Bockris, J. O., **Interpretation of Current-Potential Relationships Across Biological Cell Membranes,** in *Journal of Bioelectricity,* 1 (2) 289-294, (Marcel Dekker, Inc, 1982).

[14] study summarized in *Brain/Mind Bulletin* (Los Angeles, CA: Marilyn Ferguson, ed.).

[15] **Effects of Low Plus Lenses on Reading and Physiological Indicators of Stress,** Ph.D. thesis at Indiana University.

[16] Mancuso, T.F., Stewart, A.M., Kneale, G.W., **Radiation Exposures of Hanford Workers Dying from Cancer and Other Causes,** *Health Physics* 33: 369-384 (1977).

[17] Brookshire, D., **The Hidden Effects of Geopathic Disturbances,** (H.R.F. International, 1990).

[18] For more symptoms of these imbalances, see Table 5.

[19] Scott-Morely, A., **Geopathic Stress: the Reason Why Therapies Fail?,** in *Journal of Bioenergetic Medicine,* 3/5 (1985) 18.

➢ A young professional man in New York came to the author in 1988, suffering from constant, chronic sinus headaches and the beginnings of a serious degenerative eye condition called keratoconus. When BER testing revealed electromagnetic stress as a contributing cause, the client couldn't think of anything electrical in his bedroom, until he remembered that the fuse boxes for both his and a neighboring apartment were on either side of the head of his bed. Sleeping in another room in the house led to an immediate healing crisis. This was marked by an acute exacerbation of his chronic sinus headache, followed the next day by steady improvement for the first time in the years since he had begun sleeping in the room with the fuse boxes. He also noticed that he began remembering his dreams since moving away from the electromagnetically stressed bedroom. The side benefit of improved circulation to the head and eyes has eliminated one of the risk factors this young man once carried with him every day for the potential loss of vision. Keratoconus can and does cause blindness when it progresses.

❖

In order to deal with the effects of electromagnetic pollution in our daily lives, we must first be able to identify the sources which may be having negative effects on our bodies, and determine by their relative strength how adversely each source may be affecting us. This chapter deals with some methods of detecting the source and strength of electromagnetic fields. It is not meant to be a comprehensive listing of all available measuring devices for these purposes. The instruments referred to have varying degrees of effectiveness and ease or difficulty of use. Of those listed, those marked with an asterisk (*) are most highly recommended for use by the average person.

Detection of Electromagnetic Pollution

DIRECT MEASUREMENT DEVICES

IONIZING RADIATION

Some of the potential sources for ionizing radiation include *radon gas, sunlight, cosmic radiation, bedrock, ceramics with uranium oxide glazing, radium dial clocks and watches, lantern mantles, rock collections, leaking ionization-type smoke detectors, leaking compact fluorescent lamp ballasts, x-rays, airline travel, nuclear reactors and*

Ionizing radiation is generated by a great many sources with which we are constantly in contact.

Allowable levels of ionizing radiation have been decreased time after time as more of the dangers have become known.

nuclear wastes from medicine, industry or the military. Allowable levels have been decreased time after time as more of the dangers became known, as shown below. It is most likely that the same will be true for non-ionizing radiation some day. The following limits were set on permissible exposure for radiation workers:

Prior to 1925: no limit
1925: 46 rem per year
1934: 32 rem per year
1950: 15 rem per year
1957: 5 rem per year

For the general public, more stringent exposure limits were set in the past 4 decades:

1952: 1.5 rem per year
1958: 0.5 rem per year
(for total population)
1958: 0.17 rem per year
(average for an individual)
At present: while the standards of the 1950's remain in force, it is acknowledged that there is no level that is 100% safe, since even a single radioactive particle can cause a genetic change that could initiate cancer. This is actually happening constantly in the body, but the unstressed immune system is able to identify and eliminate these damaged cells. Increased rates of cancer and other disease have been demonstrated at levels at least 30 times lower than the levels permitted.[1]

* Single use radiation test kits for radon are now widely available for reasonable prices. This is important because radon accounts for about a third of all the background radiation we are exposed to. This simple test is needed tell you if your house is high or low in radon, since it can vary with construction and geology, even within the same neighborhood. One such test is the Quick-Screen Radon Gas Detector.[2] The least expensive test in a list compiled by Consumer Reports was only $12.[3] This is an important and inexpensive test. High radon levels have been found in homes in all 50 states.

Monitor 4 and similar dosimeters[4,5] are continuously-monitoring geiger counters for detection of nuclear radiation. Monitor 4 has a range from 0.5 to 50 mR/hour and it provides either visual or auditory count indication.

Radalert[6] is an accumulating radiation monitor which can be hooked up to IBM compatible computers for graphing and data analysis. It can also be set to produce an alert signal when a given radiation threshold is reached. It is sensitive to alpha, beta, gamma and x-rays.

Personal Radiation Detectors[7] do not measure the level of radiation, but simply "glow in the dark" in the presence of radiation levels above .4 rem per hour. They are relatively inexpensive.

Detection and measuring devices for ionizing radiation.

* **Ultraviolet light detectors** are also available. Many eye doctors have them in their offices to check UV absorption of ophthalmic lenses. Some, like the UV Sensometer, are as simple as a card that fits into your wallet.[8] If you use sunglasses, you should have one of these simple, inexpensive cards. Normally, the pupil of the eye gets smaller in bright light, protecting the inside of the eye from excess UV exposure. Sunglass tints absorb visible light, so the pupil stays larger. If these same lenses do not absorb UV, that means the sunglasses are causing an increase in UV exposure to the delicate tissues of the eye!

NATURAL FIELDS

Detection and measuring devices for naturally-occurring electro-magnetic fields.

Metal detector: A simple metal detector will detect some of the shifts in the magnetic field, such as those caused by the presence of ferromagnetic materials including iron. An example is the magnetic field aberrations induced in many beds by the metal springs of the mattress or the metal cross braces supporting the mattress.

Geo-Magnetometer: Developed by Dr. Ludger Mersmann, this sensitive instrument produces a 3 dimensional graph of geomagnetic disturbance fields.

Surface tension: The effect on the surface tension of water placed in an electromagnetic field can be measured using a sensitive balance and a platinum-iridium ring as in the Du Nouy type tensiometer.[9,10]

MAN-MADE FIELDS

SIMULTANEOUS MEASUREMENT OF ELECTRICAL AND MAGENETIC COMPONENTS:

* **Trifield Meter**[11] is an inexpensive meter recently introduced to provide independent measurement of low frequency electrical and magnetic field components. In addition, it has a setting to measure higher frequency radiation in the radiowave and microwave range.

EmdexC[12] measures all 3 magnetic field axes and the electrical field simultaneously. It is designed to be connected to a personal computer in order to provide data analysis and graphics.

Other instruments for both electrical and magnetic field measurement are available; however, they are also designed and priced for the professional.[13]

MAGNETIC COMPONENT:

* **The Trifield Meter**[14] measures the magnetic component of EMF, as well as the electrical component and radio/microwaves.

M116 Digital Magnetic Field Meter[15] is a magnetic sensor attachment for standard

Detection and measuring devices electrical and magnetic components.

multimeters to allow measurement of fields greater than 0.1 milligauss.

Magalert 660[16] measures 60 Hz magnetic fields from 0.1 to 100 milligauss.

60 Hz Magnetic Field Meter[17] measures 60 Hz magnetic fields ranging from 0.01 to 2,000 milligauss. It provides both auditory and visual alarms. Options are available for 3-axis measurements and for data logging.

Other instruments are available and are designed for the homeowner.[18]

You can even make your own simple magnetic field detector using inexpensive materials and a telephone speaker.[19]

Microwave radiation now comes at us in many forms.

MICROWAVE AND RADIO FREQUENCIES (RF):

Microwave radiation now comes at us in many forms:

alarm systems

cellular phones, CB's, pagers, walkie-talkies and other radio communication systems

diathermy and other medical uses

electronic games

remote control devices

microwave ovens

radar devices

satellite dish antennae

signal generators

The effects of microwave exposure can include:

birth defects and stillbirths

skin burns and cataracts

central nervous system effects, including dizziness, calcium mishandling, blood-brain barrier disturbance, impaired judgement and irritability

endocrine dysfunction

fatigue and muscular weakness

genetic damage

headache

cardiovascular damage

increased lymphocyte count and leukemia[20]

Detection and measuring devices for microwave radiation.

* **The Trifield Meter**[21] has a setting for measuring the radiation in the radiowave and microwave range, in addition to measuring the magnetic and electrical components of lower frequency electromagnetic radiation. Microwaves can also be detected (but not measured) by inexpensive devices available at Radio Shack stores.

AM radio: An ordinary AM radio is your most accessible device to demonstrate the presence of some types of man-made electromagnetic pollution. Some fields over about 1 milligauss will be heard as static with the volume turned up and the tuner set between stations. Fields of this strength were implicated in increasing leukemia rates in a recent Denver study[9]. Think of this the next time you experience difficulty getting good reception in part of the AM radio dial. You can experiment with your AM radio and discover some of the high field emitters in your area, such as electric rail lines, etc. To detect many EMF sources, though, the radio technique is not adequate. A meter is the way to go.

Detection and measuring devices for VLF frequencies.

VLF FREQUENCIES:

VLF frequencies are extremely important for anyone who works at a computer or sits close when watching television.

* TV Pet[23] is the simplest VLF detector on the market. It is also the most fun to use, providing a colored light feedback when located in low, borderline, and high field levels. With this device kept near the television set or the home or school computer, even children can quickly learn exactly how far to stay away from these VLF sources. This is especially important since children are even more susceptible to damage from radiation than are adults.

VLF Safe Meter[24] measures the VLF components of VDTs, television sets and some fluorescent lights in the 10^4 Hz range. It is a professional quality instrument, and its use does require some understanding of geometry and mathematics.

ELF Frequencies:

ELF frequencies are the most common source of electromagnetic stress, coming from power lines, transformers, electric wiring and appliances, including TVs and computers which also emit in the VLF range.

> Detection and measuring devices for ELF frequencies.

* **Power Pet**[25] is a simple, fun ELF detector, identical in use to the TV pet described above, which covers the VLF frequency range.

ELF Safe Meter[26] measures the 50 to 60 Hz (ELF) fields given off by the electrical distribution system and electrical appliances, including the ELF component of VDT and television emissions.

Elf Sense[27] is being designed to measure fields with frequencies in the 50 Hz to 1000 Hz frequency range. Many countries use 50 Hz electrical systems.

DETECTING BIOLOGICAL EFFECTS

Sensitive Plant Species:

> Biological monitoring of electromagnetic radiation using plants.

Mustard Seedlings: Bioassay of mustard seedlings has been used extensively to study the effects of subtle levels of non-ionizing electromagnetic radiation. Sprouts of mung bean, wheat, barley, oats, mustard, kale, rape, parsley, celery, carrot, peas, broad beans, sweet peas, lettuce and cress have also been used successfully. Plant species visibly affected by non-ionizing radiation include **mistletoe, crow garlic** (*Allium vinealis*), **yew, hawthorn, hazel, and apple**[28]. Others have been reported as well[29].

Violets (Vi*ola odorata*) and other species sensitive to ionizing radiation are used to detect the presence of radioactivity.

Kirlian photographs[30] of leaves and needles of trees and plants show bioenergetic changes to electromagnetic and sound stimuli. Plants, like animals, can transduce sound into electromagnetic fields internally.

The "L-fields" or electromagnetic fields of live trees have been continuously monitored by Professor Burr of Yale University. The L-field is a measurement of the biologically produced electrical potential of an organism.[31]

Biological monitoring of electromagnetic radiation using animals.

ANIMAL STUDIES:

The fields of life, or "L-fields" of animals of many species were studied extensively by Professor Burr as well.[32]

Historically, Greeks and Romans inspected the **livers of animals** grazing on a site before considering construction of a city there. Later, in medieval times, European builders would test building sites for geopathic stress by grazing sheep there to see if they would avoid the site. They also oriented the location and design of the city in relation to the solar system, as did the Chinese. The Emperor Kuang Yu published an edict requiring that building sites be probed for underground water currents in order to avoid their negative health effects.[33,34]

Bio-Energetic Regulation (BER) Test Methods[35]

Bio-Energetic Regulatory (BER) Medicine is becoming a major new force in the medical community of Europe and around the world. The best reason for this is its effectiveness. Rather than blocking the body's expression of symptoms, this approach focuses on supporting the body's functions. Its methods are compatible with the many traditions of natural healing handed down in cultures around the world. While drawing on these time-proven solutions from nature, BER often utilizes space age technology to monitor the minute electromagnetic functions of the body. This information base provides a cybernetic or feedback system approach to the understanding of biological processes and their responses to potential interventions. By analyzing stress patterns on an information level, only those support systems from nature which are effective and well tolerated by the individual at that time are actually used. This eliminates the waste of time and biological energy involved in trial and error approaches. Since this method taps into the biological information system which controls every body function,

Bio-Energetic Regulatory (BER) Medicine is becoming a major new force in the medical community of Europe and around the world.

even subtle but significant functional stresses of the the mind, emotions, and factors like electromagnetic stresses can be probed and neutralized before they cause greater disturbances. BER methods have evolved from the application of modern electronics to traditional healing methods such as acupuncture, homeopathy and herbal medicine.

TRADITIONAL SENSORY DETECTION SYSTEMS:

Oriental Medicine or Traditional Chinese Medicine (TCM):

Pulse diagnosis: The pulse is carried by the circulatory system, which is usually one of the first to be affected by electromagnetic stress. The pulse is observed by finger pressure at the points known as:

mu points (abdomen),

shu points (paraspinal) and

yuan (wrist and ankle) points.

Akabane is another way to find imbalances in the body's electromagnetic channels or meridians. It is a technique for measuring thermal sensitivity to a 450° F infrared stimulus at the jing well points (by the fingernails). This is an electromagnetic stimulus (infrared photons) testing the sensitivity to pain from heat. It is greatly dependent upon the quality of the circulation in its ability to dissipate that heat.

Traditional sensory detection systems which may be effective in determining the effects of electromagnetic fields.

Circulation is one of the primary functions interfered with by electromagnetic stresses.

ELECTRONIC MEASUREMENT SYSTEMS:

Burr's L-field measurements record the body's DC voltage up to about 20 millivolts at an electrode in contact with part of the body, in relation to a reference electrode elsewhere on the body. Field voltages at short distances from the body can also be measured sometimes.[36]

Ryodoraku electrodiagnosis using 6 volt, 12 volt and 21 volt DC current at the yuan points of the wrist and ankle.

Electroacupuncture According to Voll (EAV) using 0.9 volt DC at over 800 points of the EAV system, especially Voll's 40 Control Measurement Points (CMP) of the fingers and toes.

Bioelectronic Functions Diagnosis (BFD) using 0.9 volts to 5 volts DC at the Voll CMP or the fingertip jing well points.

Vegatest Method (VTM) using 1.5 volts DC at any reactive acupoint, monitoring stress and relaxation responses to electromagnetic stimuli. Such stimuli can carry information relative to organs, meridians, nutrients, foods, remedies, health conditions, toxins, chakras, colors, electromagnetic frequencies, emotions, etc.

Electronic measurement systems which may be effective in determining the effects of electromagnetic fields.

KINESIOLOGY (MUSCLE TESTING):

Applied Kinesiology (AK) Touch for Health, Educational Kinesiology (EK) and Behavioral Kinesiology (BK) systems can be utilitzed to test stresses and their relaxation through remedies, affirmations, visualizations, and other modalities.

Vega Biokinesiology (VBK) utilitzes the same procedural protocols and stimuli as the Vegatest Method, above, but monitors responses through kinesiology rather than electroacupuncture. A brief course in this method is available for primary caregivers through AERAI, as is a full practitioner certification course covering both VTM and VBK.

Life Energy Activation Program (LEAP)

Surrogate Vega Biokinesiology (SVBK) is an investigative research evaluation available to AERAI members participating in a comprehensive stress-elimination research protocol. The program is called LEAP, which stands for Life Energy Activation Program. This includes study of geopathic and electromagnetic stresses as well as the normal electromagnetic functions of the body such as the meridians and chakras.

LEAP is now available world-wide. This is due to the advent of the surrogate testing protocol, now proven successful for more than a year with clients spanning the United States. We can perform the full Vegatest analysis for those who prefer or require home-based support anywhere in the world. Following is a sample of unsolicited comments received from participants in this program. All underlines were made by the clients themselves.

M.Z. in Washington State, in response to 'blind' testing of herbal samples she had sent along with her regular testing, writes: *"The results on my test packets was really interesting. "A" was an herbal mix that is supposed to be for rebuilding of body systems. "B" was a blood cleaning formula and "C" is red raspberry leaf collected in the wild near our Idaho home. The really interesting aspect is that I wanted to start taking the teas again after I ran out of the vitamins so I used my gold ring as a pendulum to test them myself. I got the same results you did without the details. Gives me a bit more confidence in my own intuitive processes."*

P.L. in Vermont, regarding 3 family members in the program, writes, *"I see improvement here in many areas which is encouraging. Are you coming east this summer or shall we continue with hair samples?"* Three months later, the same client writes: *"I found my test results fascinating as usual and was most amazed to see the reference to my injured cervical vertebrae, which had been bothering me to extremes within the last couple of months! (for which discomfort I am off to a Shiatsu masseur) And my eyes periodically play tricks on me, neither of which symptom I mentioned. I am impressed. . . . It's fascinating to be able to actually feel myself get in and out of whack!"* The last reference related to waiting longer than optimal between follow-ups, in order to budget for participation by all family members.

A few other typical examples help to tell the story of the LEAP program:

V.W. in Washington State reported by phone today that her jaw had stopped aching until she ran out of the homeopathic "Acute Virotox." She said that she thinks the surrogate test results were correct that the jaw pain is related to Herpes zoster, not only because of the response with the support of the remedy, but also because she has had a history of Herpes zoster breaking out on the cheek in that very area in the past. She had never mentioned this part of her medical history to us before.

K.S. in New York had finished replacing all of her mercury based amalgam fillings several months previously, when testing again showed a mercury resonance. This time, however, it was at a much higher homeopathic potency, indicating that this effect had most likely been inherited or aquired *in utero*. When she discussed this test finding with her mother, K.S. learned for the first time that her mother had received several mercury based fillings during her pregnancy with K.S. Since mercury is known to cross the placenta, preventively oriented nations like Sweden have now outlawed the use of silver amalgam fillings, which contain 40 to 50% mercury, in pregnant women.

< CHAPTER III ❖ DETECTION OF ELECTROMAGNETIC POLLUTION >

Kirlian photography shows the body's own electromagnetic energy fields surrounding the fingertips and toes, which represent the endpoints of the meridians. The interpretation of these fields has been researched in Europe.[37] The high frequency electrical field produced by a Tesla coil utilized to create such a photograph is itself a powerful electromagnetic stimulus. Other methods of probing will not be as accurate immediately following such a test. It is still an excellent objective screening tool, although it should not be used by extremely electromagnetically sensitive individuals.

Segmental Electrograph (SEG) measures the body's regulatory capacity in response to varying degrees of electromagnetic stress in each of the main body areas. In this way a completely objective analysis can be made which pinpoints physical problem areas long before they can be seen with standard medical imaging techniques.

Bioelectronics of Vincent (BEV) is another totally objective assessment of the body's electromagnetic status, but in this case based upon a very precise determination of the biophysical measurement of both the electrical factor and the magnetic factor in the various body fluids. The fluids usually tested are the blood representing what the body contains, the urine representing what the body is eliminating, and the saliva representing what the body is producing in order to digest and absorb nutrients.

Additional electronic measurement systems which may be effective in determining the effects of electromagnetic fields.

Intuition, psychic perception, and direct sensations due to hypersensitivity: Each individual has a unique threshold sensitivity pattern for electromagnetic stimuli. In practice, some individuals can detect their responses to extremely minute fields. This is theoretically possible if in these extreme cases, the whole body can act as a resonator or antenna. Also the mode of perception varies. One person may notice strange feelings, another experiences ringing or buzzing in the ears, while yet a third person suffers visual disturbances from the same field. In general, hypersensitivity of the senses and of intuition are commonly seen in those who suffer from hypersensitivity to EMF. Those who are not sensitive enough to perceive effects of typical EMFs directly can often learn to amplify subliminal effects using some simple devices and techniques introduced below.

Amplification of Body Knowledge or Intuition: This is a type of biofeedback setup where subtle movement patterns are naturally amplified to a level where they can be perceived by the normal senses. Some of these methods have been practiced for many generations in various parts of the world. They continue to provide valuable information about subtle levels of energy, such as those distinguishing underground water. A good water dowser can save most of their clients thousands of dollars by finding a good water supply on the first drilling for a well.

* Metal or wood dowsing rods: Rods are designed to swing very freely and therefore to be sensitive amplifyers of slight neuro-muscular responses in the body.[38]

The Hartmann Society in Europe produces adjustable rods to pick up specific fields. By adjusting the length of the rod, it is found that specific frequencies can be more precisely scanned for.

* Pendulums: Stone or metal pendulums on the end of a cord or chain are often used to amplify the body's subliminal physiological responses to questions.[39,40] As with rods, the length of the pendulum can be varied to increase sensitivity to harmonic frequency bands.

Radionics is a form of biofeedback or amplification of subtle body responses. Radionics instruments range from home-made models to the SE5 computer. Their common element is the use of numbers or rates for the different resonances and a stick plate which is rubbed by the fingertips. The degree of stickiness felt or the sound heard when the fingers begin to stick on the material provides the amplified feedback for the subtle autonomic nervous system responses being monitored.[41,42,43]

FOOTNOTES

[1] Morgan, K.J., **Cancer and Low-Level Ionizing Radiation,** *Bulletin of the Atomic Scientists,* September (1978).

[2] $25 from Baubiologie Hardware at (800) 441-8971.

[3] $12 from Radon Project, P.O. Box 90069, Pittsburgh, PA 15224, phone (412) 687-3393.

[4] $309 from Real Goods at (800) 762-7325.

[5] Dosimeter Corporation at (800) 322-8258, or Victoreen, Inc. at (216) 795-8200

[6] $295 plus $139 for the computer cable and software from Real Goods at (800) 762-7325.

[7] Available from Micon at (417) 623-7083.

[8] $3.99 from Baubiologie Hardware at (800) 441-8971.

[9] Miller, R.N., **Methods of Detecting and Measuring Healing Energies,** in <u>Future Science</u>, White and Krippner, Ed. (Anchor Books, 1977).

[10] Fidler, J.H., **Earth Energy, A Dowser's Investigation of Ley Lines,** p. 107, (Wellingborough, Northamptonshire, Great Britain: The Aquarian Press, 1988).

[11] $100 from AERAI at (800) 788-2442.

[12] $2000 from Electric Field Measurements, P.O. Box 326, Route 183, West Stockbridge, MA 01266, phone (413) 637-1929.

[13] Write to Narda Microwave Products, 435 Moreland Road, Hauppauge, NY 11788, and to Haladay Industries, 14825 Martin Drive, Eden Prairie, MN 55344 for information on more sophisticated equipment.

[14] $100 from AERAI at (800) 788-2442.

[15] $77 from Electric Field Measurements, P.O. Box 326, Route 183, West Stockbridge, MA 01266, phone (413) 637-1929.

[16] $115 from Real Goods at (800) 762-7325.

[17] $595 plus $119 for 3-way switching probe from Environmental Testing and Technology, Inc., P.O. Box 369, Encinitas, CA 92024, phone (619) 436-5990.

[18] Write to Monitor Industries, Salina Star Route, Boulder, CO 80302, and to Integrity Electronics and Research, 558 Breckenridge Street, Buffalo, NY 14222 for information on more simple 60 Hz magnetic field monitors.

[19] Instructions are given in the video, **Current Switch**, which is $50 from AERAI at (800)788-2442.

[20] Shannon, S., **Diet for the Atomic Age, How to Protect Yourself from Low-Level Radiation**, (Wayne, NJ: Avery Publishing Group, Inc., 1987).

[21] $100 from AERAI at (800) 788-2442.

[22] **Alternative Energy Sourcebook**, (Ukiah, CO: Real Goods, 1990).

[23] $50 from AERAI at (800) 788-2442.

[24] $195 from AERAI at (800) 788-2442.

[25] $50 from AERAI at (800) 788-2442.

[26] $195 from AERAI at (800) 788-2442.

[27] Being developed by ExpanTest, Inc., 232 Saint John Street, Suite 316, Portland, ME 04102, phone (207) 871-0224.

[28] Fidler, J.H., **Earth Energy, A Dowser's Investigation of Ley Lines**, p. 97 ff, (Wellingborough, Northamptonshire, Great Britain: The Aquarian Press, 1988).

[29] Underwood, G., **The Pattern of the Past**, (Great Britain: Museum Press, 1969).

[30] Maby, J.C., and Franklin, T.B., **The Physics of the Dowsing Rod**, (Bell, 1949).

[31] Burr, H.S., **Blueprint for Immortality, The Electric Patterns of Life**, (Great Brittain: The C.W. Daniel Co. Ltd., 1972).

[32] i.b.i.d.

[33] Rossbach, S., **Feng Shui, The Chinese Art of Placement**, p. 15, (New York, NY: Dutton, 1983).

[34] Coghill, R., **Electro Pollution, How to Protect Yourself Against It**, (Wellingborough, Northamptonshire, England: Thorson's, 1990).

[35] Van Benschoten, M.M., Clinical Research in Complementray Medicine, (San Francisco, CA: Occidental Institute Research Foundation, 1989).

[36] Burr, H.S., **Blueprint for Immortality, The Electric Patterns of Life,** (Great Brittain: The C.W. Daniel Co. Ltd., 1972).

[37] Mandel, P., Energy Emission Analysis, New Application of Kirlian Photography for Holistic Health, (Essen, W. Germany: Synthesis Publishing Company).

[38] Dowsing rods, a pendulum and instructions are available as a set for $45 from AERAI at (800) 788-2442.

[39] ibid.

[40] Askew, S. and an anonymous doctor, **How to Use the Pendulum, & Diagnostic Analysis with the Pendulum,** (Mokelumne Hill, CA: Health Research, 1955).

[41] **Biofield Research Manual,** (Human Services Development Center, 1987).

[42] Tansley, D.V., **Dimensions of Radionics, New Techniques of Instrumented Distant-Healing,** (Essex, England: The C.W. Daniel Company, Ltd., 1977).

[43] Tomlinson, H., **Medical Divination, The Theory and Practice of Radiesthesia,** (Devon, England: Health Science Press, 1966).

➢ A young boy came to me as a client in New York in 1987. He was showing early body energy patterns of auto-immune reactions in the area of the rectum. BER findings initially indicated emotional stresses related to this pattern. This was confirmed by further history taking during which the mother then revealed a previously undisclosed history of sexual abuse. This stress pattern was found to match closely with the energy resonance of a particular flower essence which was then used as a remedy. The auto-immune energy pattern decreased, but was not eliminated completely at a follow-up visit a month later. Instead, a second contributing cause was now able to be identified. This was electromagnetic stress, specifically potentiating the autoimmune energy pattern in this same rectal tissue. Further discussion with the boy and his mother revealed that he would habitually climb up and sit on a metal electrical box on the roadside while waiting for the school bus each morning. By understanding the preventive significance of changing this behavior pattern, the boy was able to change his habit. Through this early intervention, perhaps years before the potential appearance of any physical pathology, he was able to eliminate this health risk pattern.

❖

Successful solutions to the subtle and complex problems of energy fields often involve a combination of several strategies. When the fields are man-made, they can potentially be eliminated or moved. When they are naturally occurring, however, experience has shown that avoidance is the best long term strategy. In any case the sensitive or susceptible individual will want to support their own bodily systems in the face of any such stressor. Support continues to be a clinically important factor in recovery from more severe stress effects for as long as several years following actual elimination of exposure to the stressor. The most commonly helpful solutions to electromagnetic pollution are discussed in this section.

Solutions to the Problem

In Traditional Chinese Medicine (TCM), the earliest known system of solutions was devised for dealing with electromagnetic stress problems:

Feng shui is the term for patterns of disharmony caused by geopathic stress. Some acupuncture practitioners are trained in its diagnosis and treatment according to traditional oriental methods. These include the nine basic cures of feng shui, all of which, as you will see when you have digested the companion volume to this book, have a direct relationship to the total electromagnetic field experienced in the human body. These nine

Feng shui is the earliest known system of solutions devised for dealing with electro-magnetic stress problems.

In addition to these historically proven approaches from oriental medicine, there are many modern strategies for eliminating, or at least reducing, the harmful effects of electro-magnetic stress.

factors, as outlined in Table 3, may be arranged to minimize stress.[1,2]

In addition to these historically proven approaches from oriental medicine, there are many modern strategies for eliminating, or at least reducing, the harmful effects of electromagnetic stress.

Helpfulness Ratings given in this section are based upon clinical experience with many hundreds of patients with electromagnetic stress problems. They are intended to provide a relative guide as to the probable helpfulness of each solution, based on such factors as cost and effectiveness for common or typical electromagnetic stress conditions. The range is from 10 (best) to 1 (least likely to be helpful). These ratings follow the possible solution, as in the following example: **Bedroom EMF** *(10):*

MOVE OR ELIMINATE THE FIELD

Man-made fields:

Bedroom EMF *(10):* This is usually the main source of the problem and usually the easiest to manage. You simply unplug and move all the electrical equipment away from your sleep place.

Radiation *(8):* If you live downstream or downwind from a nuclear installation or are exposed to low levels of radioactive

Table 3.

Nine Cures of Feng Shui

1: **Light** (e.g. mirror, crystal ball, lights):
   Light is an electromagnetic field (EMF).

2: **Sound** (wind chimes, bells, music):
   Sound produces EMF in the body.

3: **Living objects** (plants, flowers, aquarium):
   Living tissues emit physiological EMFs and absorb electrical fields.

4: **Moving objects** (mobile, windmill, fountain):
   Moving objects produce oscillating EMFs.

5: **Heavy objects** (stones, statues):
   Heavy objects absorb EMFs and often have piezoelectric properties due to their mineral content.

6: **Electrically powered objects** (air conditioner, stereo, TV):
   These emit 60 Hz EMF (or 50 Hz in many countries).

7: **Symbols of peace and protection:**
   Meaningful symbols affect the body's own EMF through imagery

8: **Colors:**
   Colors are composed of specific EMF frequencies.

9: **Others** (e.g. traditional Chinese treatments like colored ornamentation, chalk under the bed):
   These are materials which reflect or absorb certain EMF frequencies.

NASA researchers have achieved "impressive results in the use of alligator-weed, cattail reeds, and tomatoes in the de-con-tamination of radioactive soils and water."

pollution, you will be happy to know that NASA researchers Rebecca McCaleb and B.C. Wolverton have achieved "impressive results in the use of alligatorweed, cattail reeds, and tomatoes in the de-contamination of radioactive soils and water."[3] Since Chernobyl, Soviet researchers have considered using Lupines in contaminated areas, because they pull radioactive elements out of the soil. Disposal of the plants would have been the next problem to solve, so the final solution proposed was to irrigate the land with a Calcium solution. The Calcium binds the radioactive minerals, as it would in the human digestive system, and carries them down into the ground water, below the root systems of many crop plants.[4]

Work place (3): EMF exposure on the job often comes from wiring, fluorescent lights, VDTs and other electrical equipment. Make requests for needed changes and educate those in decision-making positions. Talking with co-workers about the problem may add strength in numbers, when others are experiencing electromagnetic stresses, too. You may want to circulate this book for others in your workplace to read. The helpfulness of this strategy will increase soon with more awareness, acceptance and understanding of the problem. It is in the employer's interest, after all, to have healthy, happy, sharp, energetic employees who can achieve excellence in their work

EMF exposure on the job often comes from wiring, fluorescent lights, VDTs and other electrical equipment.

through the design of a low stress work environment.

ELF Armor[5] *(10):* Installing this special magnetic field absorbing metal shield around the cathode end of the CRT reduces the magnetic field by about 77%. This allows most computers to meet the Swedish standards for magnetic field exposure. In many offices, this can benefit several people, since the magnetic field generally extends several feet in all directions from a CRT.

NoRad DB60 Computer Shield[6] *(4):* This well designed shield is grounded to prevent buildup of static electrical charges, while absorbing 99% of the electrical portion of the EMF. In addition, it reduces visual stress by eliminating glare and reflections, thus increasing contrast. This is an ideal shield when combined with a magnetic shield like ELF Armor, since the magnetic field is the one factor a shield over the screen can't eliminate. The magnetic field just happens to be the single most important factor for health, so it certainly needs to be addressed.

Many computer shields *(1):* Unfortunately, at best, most computer radiation screens only block a portion of the electrical aspect of EMF emitted by monitors using leaded acrylic.[7] It is the magnetic portion, which goes right through the screens and through the body, which is primarily implicated in causing health problems. Of course, you may still want a glare shield for your computer to reduce visual stress, and if the glare shield also blocks some of the electrical field, that is still desirable. The NoRad, above, appears to be the best option for the screen at this time.

Liquid Crystal Display (LCD) computer monitors *(4)*: These emit lower field levels than the standard cathode ray tube (CRT) monitors. If they are used on the lap, as the name "lap top" computer suggests, however, they actually produce more radiation exposure than the CRTs because of the proximity to the body.

Radiation-free monitors[8] *(10)*: This is the only way to work 100% safely with a computer from the EMF perspective. The information being processed in a normal computer can be detected and electronically decoded up to a quarter mile away. Thus, for obvious national security reasons, such a radiation-free computer had to be developed. This feat is produced by a combination of low voltage components, shields to block electrical fields, special wiring geometries to make fields cancel each other out, and Mu metal which blocks magnetic fields. These "Safe Monitors" are being produced by Safe Computing Company and are available to the general public. Of course they are already in use by places like General Electric, the Congress, the E.P.A., Yale Medical School, Harvard Medical School, New York State (which did a $5 million dollar study of the health effects of magnetic fields), Cray Research, Intel, and United Auto Workers Union. They are recommended by radiologists, general physicians, optometrists and physicists, among others. Computer programmers report a 20 to 40% increase in speed and productivity a week after switching to these monitors.

Natural fields *(0):* Moving naturally existing fields is possible, but only for a few months with existing methods. Since the fields tend to come back, this is not an effective long term solution. Since alot of time, energy and money can be put into advice and devices to move the fields, it is not generally desirable as a short term approach. It is sometimes necessary in extreme cases where no other immediate solution is effective alone, and the person cannot be moved out of the field.

MOVE THE PERSON

Man-made fields: The Office of Technology Assessment of the Congress of the United States recommends a policy of "prudent avoidance."[9]

Sleep place *(3):* It is usually not necessary to move your bed due to man-made fields. It is easier to simply unplug and move all electrical devices away from the bed in

The Office of Technology Assessment of the Congress of the United States recommends a policy of "prudent avoidance" of man-made electro-magnetic fields.

most cases. The exception is when your bed is located too near a fuse box, high current power line or other relatively immovable EMF source. In the case of high tension lines and substations, it is even necessary occasionally to move the individual to a different house entirely. This is because an entire house, or even an entire neighborhood, can be within the stress field.

Television *(10):* Time Magazine recommends sitting "at least 10 feet from the television set." Also, by reducing the number of hours you watch per day, you decrease the chronicity of the exposure, and therefore, the associated risk goes down as well. Color television sets emit more EMF than do black and white sets, yet even the black and white t.v.s have been associated with 2.5 times the risk of leukemia among children.[10]

Work place *(9):* Moving to a different work station away from the source or sources of electromagnetic pollution is often the only option available in the work place at present.

Natural fields *(10):* Moving the individual out of the way of harmful electromagnetic fields is the best solution in most cases. On their own, these fields tend to be fairly stable, since they are based on earth structures and resonances which do not typically change much on the human time scale.

Support the Body Energy System

The following supportive measures are compatible with the above means of eliminating the source of exposure to electromagnetic stress. The most effective approach usually combines eliminating the source, with specific short term support of the body energy system.

DEVICES

Many devices have been developed to support the body's natural defense mechanisms against stresses of all kinds. Some have even been specifically designed to combat electromagnetic stress. The right ones for you will depend on factors like your health, the kinds of EMF you are exposed to, and your budget. It may be beneficial to use more than one of the following devices:

Natural fiber clothing (10): This is a worthwhile investment for anyone.

Cotton futon matress on a wood frame bed (10): This eliminates a common source of electromagnetic field variation in the sleeping place. Bed springs and metallic frame components should be avoided.

Negative ion generators (10): These provide a higher concentration of beneficial negative ions in the air. Harmful positive

Many devices have been developed to support the body's natural defense mechanisms against stresses of all kinds.

ions are removed through electromagnetic effects along with their associated particulate pollution. This includes allergens such as pollen, mold, mildew, odors, viruses, bacteria, chalk dust, hair and dander, etc. Some units allow the addition of natural extracts such as anti-fungal grapefruit seed extract.

Ott Light Systems[11] *(9):* The Ott light offers state of the art full spectrum lighting which simulates a solar spectrum. These lights are specifically designed to eliminate the electromagnetic stress caused by the imbalanced visible spectrum emitted by all standard incandescent and fluorescent lights. They also eliminate harmful frequencies outside the visible octave, by special built-in shielding and grounding. If you can't get Ott lights, you may be able to install your own shielding, using chicken wire over the whole fixture and lead tape over the starters at the ends of the bulbs, and then grounding all of this. The one limitation even with the Ott light is that the light emitted is still going on and off at 60 Hz. Because flickering light like this uses more nutrients in the retina, people with light sensitivity problems may still experience symptoms. Other support is then certainly needed. Another alternative for better indoor lighting is various balanced spectrum incandescent bulbs. Some of these use a purplish Neodymium coating inside the bulb, while others appear slightly blue. These lights do not contain the ultraviolet portion of the full solar spectrum, but do present a balanced visible spectrum of light for reduced electromagnetic stress.[12]

Polarizers[13] *(8):* Polarizers modulate and harmonize electromagnetic fields. This can dramatically reduce the degree of stress on the body systems. It is wise to combine the use of polarizers with other supports such as remedies tested specifically for the individual. Polarizers are available both as personal models worn on the body and larger units which affect an entire motor vehicle or an entire household's or office's circuitry when properly placed.

Subtle Energies[14] **electromagnetic field interrupters** *(8):* These have been producing some very positive feedback by users who are very sensitive to energy fields. They work by emitting a neutralizing field, which acts in the space around the unit, cancelling out the existing EMF through destructive interference.

12 volt direct current electrical systems *(8):* Direct current systems eliminate the 60 Hz fields which are a primary problem. Constant (non-oscillating) EMF is present with these systems however, when the current is on, although the degree of stress is much less. Note that when an inverter is added to a 12 volt system, 60 Hz oscillating current and fields are again present. Also, 12 volt fluorescent lighting produces a 120 Hz oscillation. For details on specific equipment available see catalogs such as Real

Goods Trading Company, 1-800-762-7325, or your local 12 volt dealer.

Speaker phones *(7):* By moving the microphone and speaker away from your mouth and ear, you reduce the electromagnetic stress. This is especially important if you spend alot of time on the phone.

Japan Life magnetic sleep system and other Japan Life products *(7):* These products produce a neutral polarity constant magnetic field which helps to mask other interfering electromagnetic fields. There are several commonly encountered problems with the system and its imitators. In the long term, chronic problems can still result from underlying masked electromagnetic fields. Clinical experience has shown that the most effective use of such uniform neutral magnetic field systems is on a brief daily basis. A 20 to 30 minute nap or meditation period in such a field in a location free of geopathic and electromagnetic stress is the ideal.

Electronic protection devices (e.g. belt attachment) *(7):* These units provide beneficial electromagnetic fields, without metal in contact with the body.

Teslar watch *(6)* The simplest solution to the electromagnetic stress of quartz watches is not to wear them on the body. They can be carried in a purse or brief case. If you can find a manual wind-up watch or self winding watch, these are a good solution, too, except that you are still wearing metal on the body. The metal can act as an antenna, picking up and transmitting unwanted environmental frequencies to your body. This is true of the Teslar watches as well, the difference being that the Teslar watch also specifically produces an 8 Hz oscillation which is beneficial for most people. It is similar to one of the main components of the Schumann field which is deficient in most of our indoor environments. The primary limitation of the Teslar watch is that it produces only this one frequency as compared to the broad spectrum rich in harmonics produced by the earth.

Non-heated water bed *(5):* While eliminating the severe problems associated with heating coils of most water beds, the presence of water as a conductive medium close to the body can still be a problem. It can amplify other existing electromagnetic fields.

Magnets: point magnets, bar magnets,[15] **magnet belt and magnetic jewelry** *(1):* These magnetic devices do work to stimulate the body's energy field or mask externally applied fields. Unfortunately they also can create other problems for those who still have amalgam fillings or other metal in the body. While they can be an important and effective part of therapy when used by knowledgeable professionals, they are not recommended for indiscriminate use.

Magnetic fields have powerful effects that not only vary at different frequencies, but also vary at the North and South poles of a static field.

Magnetic fields have powerful effects that not only vary at different frequencies, but also vary at the North and South poles of a static field. The poles differentially affect the pH, which is also called the magnetic factor. The North pole increases alkalinity, while the South pole increases acidity. This is the basis of static magnetic field therapies, whether using powerful bar magnets of 8,000 gauss to saturate an entire organ, or a tiny point magnet of 800 gauss to stimulate a single acupuncture point without inserting a needle. The stimulatory South pole especially should be used with caution as it can increase growth of tumors or of infectious organisms. Special caution is also indicated around the eyes and head, especially for prolonged exposure to large or high gauss fields.

FOODS[16]

In order to minimize the effects of electromagnetic pollution on your body, the foods to avoid include:

irradiated foods
processed and refined foods
fatty foods
dairy
sugar
meat

After radioactive fallout, certain other foods should be avoided:

> *green leafy vegetables,* because they catch the short term fallout

> *root vegetables and fruits from deep rooted trees,* since they accumulate more of the long-lived fallout such as Cesium and Strontium

> *fresh water fish,* since they absorb radioactive minerals easily due to the lack of other minerals that are present in sea water.

Some general EMF protective foods to emphasize in your diet include:

whole grains

fresh vegetables

beans, miso, tofu and tempeh

sea vegetables

seeds and nuts

Food factors which can bind radioactive substances in the digestive tract and prevent absorption into the body include:

calcium

pectin and other fiber found in plant foods

phytates in seeds and grains

sodium alginate in sea vegetables

sulfur-bearing amino acids in onions, garlic, beans, eggs and cruciferous (cabbage family) vegetables

zybicolin in miso

clay, e.g. bentonite

When planning your diet, it is important to consider the effects of different foods on your pH, the magnetic factor. Research shows that when the pH is balanced or "neutral," there is less effect from radiation.[17]

REMEDIES & REDUCTION OF THE SUSCEPTIBILITY FACTOR

There is nowhere on earth that one can totally avoid the electromagnetic fields of modern culture, let alone those of nature.

There is nowhere on earth that one can totally avoid the electromagnetic fields of modern culture, let alone those of nature. Because electromagnetic radiation is know to affect some people more than others, it is clearly beneficial to improve the overall performance of your biological system. This reduces your susceptibility to stress, damage and health risks. Susceptibility, for example to radiation induced leukemia, can be 12 times higher than normal for people such as:

pregnant women, infants and young children

asthmatics

people with allergies

heart patients

diabetics

arthritics

prematurely aging individuals

elderly

people with environmental sensitivities

Essentially, young children are sensitive because their bodies are actively forming. The infirm are more sensitive because they are already coping poorly with the toxins present in their systems. Energizing those toxins even with non-ionizing radiation simply speeds up the degenerative processes.

Remedies are natural support substances or energies which help to promote healthy regulation of the body's normal functions. The best remedy is always an individual matter determined not only by the stress but also by the state of the body and its resultant response to the imposed stress. The best approach therefore, is an individualized one that observes in detail the functional behavior patterns of the physiological system to be supported. Systematic approaches to this include oriental medicine, homeopathy and also modern Bio-Energetic Regulatory (BER) methods. BER incorporates many tools and concepts from the earlier approaches to functional biological observation from both East and West. It is not

The best remedy for stress induced by electro-magnetic pollution is always an individual matter determined not only by the stress but also by the state of the body and its resultant response to the imposed stress.

limited to any philosophical framework or theoretical model. This allows the necessary flexibility to best analyze the interconnecting complexities of stresses found in today's population which run the gamut from EMF to chemical pollution to psychological stresses.

Energetic, nutritional and other individualized support based on Bio-Energetic Regulatory testing *(10):* This is the best method of support because it addresses the unique problems and needs of each person. Ongoing investigational research is being performed by AERAI and participating members. Nutrients and remedies protective against radiation that can be tested for individual effectiveness and tolerance include:[18,19]

Vitamins: beta carotene, A, B6, B12, pantothenic acid, inositol, choline, PABA, bioflavonoids, C and E

Minerals: Calcium, Chromium, Copper, Iodine, Iron, Magnesium, Manganese, Potassium, Selenium and Zinc

Amino acids: Cysteine, sulfur bearing amino acids in SGP garlic or Kyolic, Histidine and the three amino acid chain Glutathione.

Enzymes: SOD, Glutathione Peroxidase.

RNA, and its breakdown product uric acid (a potent anti-oxidant).

Adaptogens: Ginseng, Eleutherococcus, Royal Jelly, Flower Pollen.

Support for elimination of radioactivity sources in the body: high fiber foods, pectin, algin, kelp, Iodine, Calcium, Magnesium, clay (bentonite), Vitamin A, C, E, beta carotene, acidophilus.

Essential Fatty Acids:

omega-3 fatty acids (alpha linolenic acid, EPA) from cold water fish like salmon, mackerel and herring.

omega-6 fatty acids (linoleic acid, gamma linolenic acid, arachidonic acid) from seeds like sunflower, sesame and pumpkin, and oils like evening primrose, safflower, sunflower and linseed.

Radiation antitox[20] (Apex Energetics) *(9):* This is an example of state of the art complex homeo-energetic formulation. Radiation antitox is designed specifically to give support to those experiencing symptoms of electromagnetic stress. This complex helps to balance all of the acupuncture meridians. In cases of chronic electromagnetic stress, it is often combined with other complexes such as those for anti-oxidant protection, and restoration of cellular energy functions. Vitamins and minerals, such as A, B complex, C, plus zinc and selenium can be helpful partners as well. This remedy works to activate the

Table 4.

Composition of Radiation Antitox

Homeopathics (electromagnetically energized dilutions of herbs, minerals and other substances):

    aloe (herb)
    echinacea angustifolia (herb)
    echinacea purpurea (herb)
    ferrum metallicum
    fucus vesiculosus (herb)
    iodum
    kali phosphoricum (also a cell salt)
    lecithina
    phosphorus (also a cell salt)

Sarcodes (homeopathically derived glandular extracts from healthy organically raised animals):

    bone marrow
    hematopoetic tissue
    placenta
    skin

Orthomolecular (nutritional) substances:

    DNA
    iodine
    lecithin
    Vitamin A
    Vitamin E

Gemmotherapy (extracted from buds or other highly acitve plant tissue, and used for the stimulation of drainage of toxins):

    Ribes nigrum gemmae

immune system and energize the cellular metabolism. Among its 52 ingredients are those listed in Table 4.

Electromagnetic Stress flower essence combination (AERAI) *(9):* The component flowers essences and their functions are given below. They can be ordered pre-mixed for personal research through AERAI.

Blackberry: overcoming energetic blockage and inertia

Dill: assimilating high frequency stimulation

Mariposa Lily: maintaining bonding with mother earth

Pink Yarrow: strengthening the body's energy field in relation to other interpenetrating electromagnetic fields

Self Heal: awakens blocked inner resources

Trillium:: clears base chakra of unrefined vibrations which trigger survival programming (including sympathetic nervous system stress response)

Detoxosode Radiation (HOBON) *(9):* This complex homeopathic remedy contains several ingredients helpful for support against ionizing radiation stresses.[21] These ingredients include:

Aloe socotrina: an herb in the lily family

Diatomaceous earth: microscopic fossils of silica

Pectins from fruits

Sea vegetation: for alginate content

Feng Shui tea *(9)* from Five Element Herb Company: This tea is a combination of traditional oriental herbs designed to help reduce the body's susceptibility to electromagnetic stresses. It is taken as a beverage over a period of several months. The constituents and their functions in oriental medicine are:

Chen Pi: regulates qi and prevents stagnation

Plantago: drains dampness and clears heat

Crataegus: for food stagnation and congealed blood

Polygonati: tonifies spleen and moistens lungs

Agastache: for dampness obstructing middle burner

Schizandra: astringent and for weak lungs and kidneys

Sophora: cools heat in liver and blood

Cortex Mori Alba: relieves cough and reduces edema

Peucedanum: relieves cough and expels phlegm

Areca: for qi stagnation and damp stagnation

Lili: moistens lung yin and calms the spirit

Perilla Seeds: regulates lungs and moistens intestines

Smokey quartz *(7)* for radiation (x-ray): The grey, smokey variety of quartz is made by the absorption of radiation by clear quartz crystals in nature over thousands or millions of years. Some of the smokey quartz crystals now on the market are produced artificially and do not have the same radiation absorption qualities. Research has investigated the use of this property in the elimination of residual electromagnetic stresses from tissues exposed to x-ray radiation, such as medical and dental x-rays. The point of the crystal is placed against the exposed area briefly until some relief in symptoms is noted. Alternatively, smokey quartz can be tested for effectiveness and compatibility in a Vegatesting type of protocol. It has been found that the crystal should be periodically exposed to red light in order to clear it for further use. In this way, effective use of the same crystal can be sustained indefinitely.[22]

Rhododendron multiple potency ampule *(4):* This combination of several homeopathic potencies (carrier frequencies) of the rhododendron herb is worn to protect against electromagnetic stress.

Arteriosclerose nosode (Staufen-Pharma) *(2):* Physicians licensed to use nosodes may prescribe an individual potency or use a potency chord. This should only be used on an individual basis by a knowledgeable practitioner on the basis of a homeopathic interview and/or a functional testing protocol.

COLOR AND LIGHT

Color and light are intimately related to electromagnetic stress. The visible octave of the electromagnetic spectrum is the one to which we are most sensitive. Our ability to receive appropriate light information and energy stimuli from our environment is a key factor in the regulation of every cell in the body. Melatonin, the hormone produced by the pineal in response to the light/dark cycle, goes to every cell in the body. When regulated properly by environmental light, it is the most potent force know for longevity and prevention of cancer. Misregulated, it can actually promote cancer. Color has been shown to shift the balance of the autonomic nervous system, which innervates every organ in the body. This dual control system consists of the parasympathetic division and the sympathetic division. This system can become imbalanced by environmental stimuli such as EMF, artificial light and other stressors. The individual can also adapt to acute and chronic stresses in an attempt to maintain homeostasis. This

Color and light are intimately related to electromagnetic stress.

Our ability to receive appropriate light information and energy stimuli from our environment is a key factor in the regulation of every cell in the body.

imbedded adaptation is another major cause of imbalance, because it interferes with the ability to respond optimally even to a stress free environment. Note the types of dysfunctions which are produced by imbalances in the sympathetic and parasympathetic nervous systems as listed in Table 5.

Color therapy *(7):* A full report on color and its uses in human health, development and performance can be obtained from AERAI for $5, and a complete set of 12 color filter combinations is also available ($20) for those wishing to practice this art for themselves or as an adjunct to their practice.

Syntonics is the optometric use of color through the eyes to improve visual health and fitness. If you experience problems with your eyes or vision that are not solved by the usual means, this is an excellent visual therapy to turn to. These frequently misunderstood problems often relate to poor or distorted ability to effectively receive the colors of the electromagnetic spectrum. Referral to the nearest qualified practitioner is available by writing to the College of Syntonic Optometry[23]. If this therapy is not yet available in your area, locate the nearest behavioral optometrist for more traditional approaches to vision therapy through the Optometric Extension Program Foundation International[24] or the College of Optometrists in Vision

Table 5. Part I

Symptoms of Autonomic Imbalance

SOME OF THE SYMPTOMS OF RELATIVE PARASYMPATHETIC, ANABOLIC OVERACTIVITY ARE:

small pupils
wide eyed
one eye may turn inward
eye strain with nausea and headache
migraine
tearing
puffy, droopy upper eyelid
low pressure in the eyes, but also glaucoma
congestion of sinuses, mucous membranes, bronchi
hay fever
excess salivation
excess hunger
excess secretion and motility of intestinal tract
    (hyperchlorhydria, pain, spastic constipation,
    diarrhea, incontinence)
enuresis, cystitis, irritable bladder
slow heart rate
low blood pressure
low blood sugar, but also diabetes
lack of sweating, eczema, urticaria
rheumatoid arthritis
low respiration rate
asthma, spasmodic laryngitis, croup
hypothyroid
excess activity of: parathyroids, adrenal cortex,
    stomach, liver, pancreas, spleen, intestines.

Table 5. Part II

Symptoms of Autonomic Imbalance

SOME SYMPTOMS OF RELATIVE SYMPATHETIC CATABOLIC OVERACTIVITY INCLUDE:

large pupils
protruding eyeballs
retinal bleeding
dry eyes
upper eyelid raised
high pressure in the eyes
poor visual focusing
eye turns outward
difficulty concentrating on close work
dry nose and throat
dry mouth
stops secretion, movement and digestion in gastro-
   intestinal tract
catarrhal gastritis, gastric ulcer
typical (atonic) constipation due to decreased peristalsis
fast pulse
high blood pressure even before arteriosclerosis
angina pectoris, myocarditis
high blood sugar
goose bumps and cold sweaty skin
increased internal body heat
perspiration from palms, soles, underarms
decreased urination, dysuria
dysmenorrhea
uterine cramps
excess activity of: thyroid, adrenal medulla, pituitary,
   gonads, muscles.

Development.[25] The traditional vision therapy approaches have been shown to enhance color receptivity and discrimination when approached as a learning process rather than an exercise.[26] Eye exercise per se has been associated in the research more with actual decreases in light receptivity and integration.

Neurosensory Development is a method of color therapy utilizing the Lumatron instrument developed by Dr. John Downing. More information on instruments and practitioner certification is available through AERAI or the Downing Institute.[27]

Spectro-chrome is a system of color therapy in which the color is applied to the body rather than to the eyes.

Magenta is often indicated on the body in cases of electromagnetic stress. It is especially beneficial over the chest and low-back. It helps the body to minimize the circulatory effects of electromagnetic stress.[28]

Night lights (3): If it is necessary to continue the use of a night light, or to use a flashlight during the night, then you should definitely use a red filter to eliminate suppression of pineal function.[29]

ACTIVITIES

Visualization: Imagine the source of stress becoming smaller and further away, while experiencing deepening breath with bodily relaxation.

Meditation: Explore a variety of methods and select one or more which suit you at this time.[30]

Vision Therapy activities: Because vision is the dominant sense for humans, and because it is also our most sensitive electro-magnetic reception system, increased development and flexibility of this system reduces the total electromagnetic stress.[31,32]

Match your activity cycle to the sun: Sunlight out of synchronicity with the biological clock's rhythm becomes a major electromagnetic stress. This is a major problem for night and swing shift workers. It is also a major element of jet lag. Keeping a regular activity cycle each day is very helpful.

If you need assistance with your personal investigation or implementation of the best combined stress elimination approach for you, write to AERAI. You do not have to be alone in your search, when others have had similar experience before. Sharing your experiences can also be valuable to other members as well as to the many people out there who haven't been able to identify the cause or solution to the stress they experience. As a member, AERAI will also provide access to appropriate support systems currently being studied.

< ELECTROMAGNETIC POLLUTION SOLUTIONS >

FOOTNOTES

[1] Rossbach, S., **Feng Shui, The Chinese Art of Placement,** (New York, NY: Dutton, 1983).

[2] Rossbach, S., **Interior Design With Feng Shui,** (New York, NY: Dutton, 1987).

[3] Rimland, L., On Constructed Wetlands and Bioremediation, in*Seeds of Change Consultancy Services & Diversity Catalogue,* 16-17 (Santa Fe, NM:Seeds of Change, 1990).

[4] Edwards, M., **Chernobyl - One Year After,** *National Geographic* 171: 633-653 (April, 1987).

[5] $79, but only available for Macintosh Plus, SE, II and Classic at the time of writing. Enquire about availability for other models. Available from AERAI at (800) 788-2442. Installation is not included and should be done by a person familiar with internal servicing of computers.

[6] $129 for flat computer housing, $139 for curved housing, and $299 for custom shield for large screens, available from Baubiologie Hardware at (800) 441-8971.

[7] Design West at (714) 833-3500, I-Protect at (213) 215-1664, Langley-St. Clair Instrumentation Systems, Inc. at (800) 221-7070, or Technograph at (416) 421-0202.

[8] Safe Computing Company, $1195 for IBM PC/XT/AT or Apple II (CGA 640 by 200 resolution), $1795 for IBM PS/2, VGAs or small MACs (VGA 640 by 480 resolution), and $1995 for Apple MAC II (VGA 640 by 480 resolution); available through AERAI at (800) 788-2442.

[9] *Business Week,* October 30, 1989.

[10] University of Southern California study of 464 children, February, 1991.

[11] Facts of Light, (Santa Barbara, CA: Ott Light Systems, 1988). Units starting at $318 are available through AERAI at (800) 788-2442.

[12] For example, Chromalux bulbs (Neodymium coated) are $9.49 for 60 watts and $9.99 for 100 watts from Baubiologie Hardware at (800) 441-8971. Balanced spectrum incandescent bulbs are also available from AERAI at (800) 788-2442.

[13] Available for $12 from AERAI at (800) 788-2442.

[14] Available from Europa Alexandra O'Tahnnt at (505) 296-4869.

[15] Broeringmeyer, R. and Broeringmeyer, M., **Energy Therapy Training Manual,** (Murray KY: Bio Health Enterprises, 1987).

[16] Shannon, S., **Diet for the Atomic Age, How to Protect Yourself from Low-Level Radiation,** (Wayne, NJ: Avery Publishing Group, Inc., 1987).

[17] Haveman, J., **The Influence of pH on the Survival after X-Irradiation of Cultured Malignant Cells,** *International Journal of Radiation Biology* 37: 201-205 (1980).

[18] Shannon, S., **Diet for the Atomic Age, How to Protect Yourself From Low-Level Radiation,** (Wayne, NJ: Avery Publishing Group, Inc., 1987).

[19] Chaitow, L., and Kutter, E., **How to Live with Low-Level Radiation, A Nutritional Protection Plan,** (Rochester, VT: Healing Arts Press, 1988).

[20] Martina, R., **Futureplex Antitox Biotherapy, Physician's Desk Reference,** (Glendale, CA: Apex Energetics, 1988).

[21] **HoBoN, Adaptosodes, Biosode O/S, Detoxosodes,** (Las Vegas, NV: HoBoN).

[22] Amonius, R., **The Use of Smokey Quartz Crystals to Clear the Effects of Radiation: Preliminary Results,** in *Psychic Research Newsletter,* V, 5, p. 5 (San Jose, CA: Psychic Research Institute, 1988).

[23] Dr. Solomon Slobins, Secretary, College of Syntonic Optometry, 1200 Robeson Street, Fall River, MA 02720.

[24] Optometric Extension Program Foundation International, 2912 South Daimler Street, Santa Ana, CA 92705-5811, phone (714) 250-8070.

[25] Dr. Robert Wold, Secretary, College of Optometrists in Vision Development, 353 H. Street, Suite C, Chula Vista, CA 92010.

[26] Swartwout, J.B. and Swartwout, G.M., Color Perception Changes Following Vision Therapy, in **Procedings of the 1988 Northeast Vision Conference** (Lebanon, OR: Caryl Croisant, 1988).

[27] Downing TechniquePractitioner Certification Training Manual, (San Francisco, CA: Downing Institute, 1989).

[28] The combination of filters which is used to make magenta is available from AERAI for $3 complete with instructions on proper use.

[29] The red filter material combination is available for $3 complete with instructions from AERAI.

[30] LeShan, L., **How to Meditate, A Guide to Self-Discovery,** (New York, NY: Bantam Books, 1974).

[31] Swartwout, J.B., **Optometric Vision Therapy, Introduction to Behavioral Optometry,** (Santa Ana, CA: Optometric Extension Program Foundation, 1977).

[32] Swartwout, G.M., **Vision Improvement** audio tape and materials, (Volcano, HI: AERAI, 1991).

➤ As another example of the kind of interactions that may take place, one patient's problems, as reported at a conference on functional medicine, did not resolve until an interaction between a watch and a ring was identified as the source. The ring had a large crystal which picked up the oscillation frequency of the quartz crystal in the watch. The metal of the ring then conducted that resonance to the two meridians on the ring finger: those for the endocrine system and for processes of organ degeneration. Several analogies can help illustrate this principal of energy transfer through resonance. One example is that if you hold a bare fluorescent light bulb, without any wires or fixture, near power transmission lines, it will actually light up. Another example that can be experienced is that if you have two stringed instruments or tuning forks which are in tune with each other, you can sound a tone on one of them, with the other a distance away. Now stop the vibration producing the tone on the first instrument. You will be able to hear and feel the vibration of the second instrument. Today, of course, we are all so familiar with electronic communications, that we almost take the concept for granted. Think for a moment, however, about your radio, your TV, your cellular phone, the satellite link for a long distance phone call or a cable TV show from a distant city, your remote controls for TV and VCR, etc. All of these common devices of the information age operate on the same principle. The first radios were crystal sets. A crystal was caused to resonate by electromagnetic radiation in the radio wave region of the electromagnetic spectrum. The rest of the spectrum, and the rest of the universe, operate no differently.

❖

Ongoing research on Electromagnetic Stresses and Solutions is being carried out by the Achievement of Excellence Research Academy International (AERAI)[1], as well as a small number of other, mostly private research organizations and clinics like the Hawaii Center for Natural Medicine[2]. Research participation at AERAI is available only to members and to clients of research associates of the Academy. More information on AERAI is avaliable for those who are interested in researching their own electromagnetic stresses or those of their clients.[1,2]

Electromagnetic Stress Research

If your entire house is subject to high magnetic field levels, it may be wise to move soon. Not only will this reduce your family's long term health risks, but you could also save some of your investment in the property. "With regard to buying a house. . . It might be prudent to consider the location of distribution and transmission power lines. . . ."[3] Already, according to Business Week, there are more than 100 lawsuits about power lines and property devaluation. Still other lawsuits, like that of a family in the state of Washington, deal with the deaths of children from leukemia and

If your entire house is subject to high magnetic field levels, it may be wise to move soon.

other cancers linked to magnetic field exposures in schools or homes. The child who died in Washington was just one of several children attending the same school that was exposed to high EMF from power lines, all of whom died within a short period of time. Beyond studying such repeated and unnecessary tragedies, it is time to begin doing what we can to prevent them, bringing to bear our present level of understanding, although it is incomplete, as it will always be. When so many children are dying daily in so many advanced nations around the world, we should not claim the luxury of waiting until further research is completed. We know now quite enough to reduce the major risk factors.

More than 60 studies have linked human exposure to magnetic fields like those emitted from power lines and computers to cancer, leukemia and tumors.

Martin Halper, a director of the Environmental Protection Agency, is quoted in Fortune Magazine saying, "In all my years of looking at chemicals, I have never seen a set of epidemiological studies that remotely approached the weight of evidence that we're seeing with ELF electromagnetic fields. Clearly there is something here."[4]

More than 60 studies have linked human exposure to magnetic fields like those emitted from power lines and computers to cancer, leukemia and tumors. In vitro laboratory studies show that human cancer cells can reproduce up to 24 times more rapidly when exposed to fields like those near power lines.[5] Based on the results of the Savitz[6] study,

Fortune magazine reports that up to 30% of all childhood cancers may be attributable to ELF fields. Dr. David Carpenter, dean of Public Health at SUNY in Albany, New York, who headed the 5 million dollar New York Power Lines Project to review the carcinogenicity of magnetic fields, calls that estimate conservative. This is a burden both in human life and in fiscal terms that must be weighed heavily against any cost for modification of our present systems for electrical power supply. If present trends continue, by the turn of the century 41% of Americans can expect to develop cancer in their lifetime. Someone dies of cancer every minute on the average in the United States. This is about a half a million deaths each year. If that figure could be reduced by one third, approximately $25 billion dollars could be redirected from cancer treatment and research to other human needs. The key is that it will first require a shared commitment as a society to take constructive preventive action.

It is no accident that most of the epidemiological studies show primarily cancers of the immune system or the nervous system in relation to electromagnetic stress, since these are the most electromagnetically active tissues in the body. No wonder they are the most disturbed by electromagnetic interference applied from outside the body. Unfortunately, the interest generated by some of these studies in the relationship between electromagnetic stress and cancer has largely failed to carry

If present trends continue, by the turn of the century 41% of Americans can expect to develop cancer in their lifetime.

It will first require a shared commitment as a society to take constructive preventive action.

Cancer only occurs in an electromagnetically low energy terrain in the body.

over into concern or study of the more subtle effects on human health and performance which must precede the onset of cancer and the ultimate death of the individual. Cancer only occurs in an electromagnetically low energy terrain in the body. In such terrain, cells lack adequate energy to fully remove metabolic wastes and environmental toxins. As these build up within cells, they further impair cellular metabolism, including energy production and detoxification. The resultant positive feedback loop is what physicians diagnose as cancer. It is simply a collection of cells that are full of toxins. As the toxins build up faster than they can be eliminated, more cells must be produced to hold them. The image is rather like a modern city dump bulging at its seams for the excess of waste material. Just as in our city dump analogy, the tumor which forms (often encapsulated, like a well managed landfill), keeps those extra poisons away from the rest of the body where they would pollute other tissues and interfere with still more links in the delicate chain of life. Many years ago, experiments in Europe in fact found a cure for cancer. Tumors were simply innoculated with a microorganism which would feed on the tumor. Unfortunately, while the disease was cured, many patients died. They died of toxicity, not of toxins produced by the microorganisms, but of the toxins being sequestered in the tumor for the protection of the rest of the body. Thus, the frequently held view of cancer cells as the enemy only leads us astray from providing the kinds of support those valiant and loyal cells need in such a depleted and toxic state. Of course, beyond a specific, irreversible level of toxin accumulation, it makes total sense to attempt to heroically excise the toxic mass from the body. What makes less sense is to attack the tumor with more toxins (chemotherapy) and radiation. Fortunately, Vincent in France has pioneered a method to measure the body's electromagnetic energy status. This can help predict whether the irreversible terrain has been enterred yet or not.

Electromagnetically based therapies for cancer have also been pioneered around the world by several researchers, with indications of good success during the reversible stages. These include the use of specific ELF electromagnetic stimuli by Rife in the U.S.A., which he found to destructively resonate with the microbes he found associated with most cancers at 666 Hz, 690 Hz and 727 Hz.[7]

Electromagnetic testing of the body's functional responses has been developed by clinical researchers in both Europe and Japan. Methods in use today on a worldwide basis include EAV,[8] BFD,[9] Vega,[10] and Ryodo Raku.[11] These and similar methods allow skilled practitioners to tap into the body's own electromagnetic regulatory system to discover its energetic imbalances, causal structure, priorities for correction, and the effectiveness

Electromagnetically based therapies for cancer have been pioneered around the world by several researchers, with indications of good success during the reversible stages.

and tolerance of specific potential therapeutic agents before they are actually instituted on a clinical trial basis. Much trial and error, and the associated stress on the body's already compromised regulatory capacity can be saved in this way in the hands of a skilled and knowledgeable clinician.

Energetic biofeedback and 'dialysis' of the body's own electromagnetic oscillations is the concept behind other methods emerging in Europe in the field of biological medicine

Energetic biofeedback and 'dialysis' of the body's own electromagnetic oscillations is the concept behind other methods emerging in Europe in the field of biological medicine. MORA therapy instruments separate harmonic, or normal, life frequencies from disharmonic elements of the body's total electromagnetic oscillation pattern. Harmonic energy can be returned to the body and amplified to an appropriately subtle degree to help support a return to a normal biological energy level from the low energy terrain which predisposes one to cancer. At the same time, the disharmonic wave forms can be inverted and amplified, so that when they are returned to the body, they destructively interfere with the sources of the disharmonic vibrations. This concept initially sounds like medicine from the next century, not from today, yet the same technology is in use for damping sound vibrations at places like Heathrow airport in England. A microphone picks up the current noise oscillations in the environment at the listeners ear, and the inverted wave form is played out through headphones. Where there was noise, now a speaker plays silence!

It is unfortunate for the 500,000 annual victims of potentially preventable death by cancer each year in the U.S.A. that no research on these techniques has been supported by the $75 billion dollar a year cancer cartel made up of the pharmaceutical industry, the medical industry and government.

Animal studies have demonstrated increased rates of birth defects and miscarriages. Three studies in peer review medical journals like the American Journal of Epidemiology link computer use with birth defects and miscarriages in humans. Three other studies link male breast cancer with magnetic fields from power lines. More than 60 studies showing health effects of EMF are summarized in the second volume of this series.[12]

Conclusion

The flow of life energy is the source of our aliveness, our wellbeing, and our consciousness. This life force is a harmonic concordance between our personal energy and that of our environment. Ultimately only one fundamental process exists in the universe. In physics it is called the superstring. It is a ray of light. Matter is merely frozen light made up of these superstrings. In each of us, the energy flow of life manifests as movement, nerve and fluid currents, emotion, and thought as well as the subtle energy fields given terms like nadis, chakras, points and meridians. When the flow is disrupted or disturbed, health cannot be achieved. When discordant or disharmonious energy fields are imposed upon us, the flow of life energy is impeded and the preconditions are set for disease. The purpose of releasing the flow of life energy is to permit the normal state of radiant health and wholeness to unfold.

Total health is a state of progressive organic development from the moment of conception to the moment of transcendence. This process encompasses physical, mental, emotional, and spiritual development. A truly healthy person achieves progress in self mastery in order to contribute uniquely to the greater good.

The functional capacity to consistently utilize a specific physiological process is an ability. Abilities encompass all spheres of physical, mental and spiritual aspects of our lives. Each person has abilities in which they are gifted and abilities that are more or less latent. These functional capacities are measurable. They can be specifically stimulated and developed. Through exploration and utilization, physiological abilities naturally develop and become an integral part of the whole person. Performance is the complex orchestration of functional capacities to achieve excellence in the creative manifestation of our goals.

For 1in 3 citizens in industrialized nations, finding solutions to electromagnetic pollution is a necessary first step toward achieving those goals.

FOOTNOTES

[1] Swartwout, G.M., **Electromagnetic Health Research Manual,** (Hilo, Hawaii: Achievement of Excellence Research Academy International, 1991).

[2] Hawaii Center for Natural Medicine, 100 Kam Ave, Hilo, HI 96720, phone (800) 788-2442.

[3] The New York Times, June 25, 1989.

[4] *Fortune Magazine*, December, 1990.

[5] Research at the Cancer Therapy and Research Foundation in San Antonio, Texas, reported in *U.S. News and World Report*, March 30, 1987.

[6] "Dr. Savitz mentioned 2 milligauss as the level at which cancer is produced in children. In the home of these people, I have measured 10 to 12 milligauss, depending upon how much current is flowing down the line." House of Representatives, 100th Congress.

[7] Lynes, B., **The Healing of Cancer, The Cures - the Cover-ups and the Solution Now!,** (Queensville, Ontario, Canada: Marcus Books, 1989).

[8] **The O.I.C.S. 'Voll' Electroaupuncture Desk Reference Manual,** (Bellingham, Washington: Occidental Institute Research Foundation, 1980).

[9] Kenyon, J.N., **Modern Techniques of Acupuncture,** (Northamptonshire, England: Thorson's Publishing, 1985).

[10] Ibid.

[11] Ibid.

[12] Swartwout, G., **Electromagnetic Health Research Manual,** (Hilo, Hawaii: AERAI, 1991).

➤ The author's father, J. Baxter Swartwout, O.D., F.C:O.V.D., F.A.A.O., a behavioral optometrist near Albany, New York, has been monitored continuously by Vegatest for several years. When he began using his electric blanket at the onset of upstate New York's chilly autumn nights in 1988, electromagnetic stress was the only test out of several hundred that came up positive. Usually a much more varied pattern of stress responses can be elicited. After five days of not using the blanket, the underlying stresses could then be identified and eliminated. In other words, the electromagnetic stress had been totally masking all other physical stresses and preventing the body from responding to them. This type of masking effect can be responsible for poor elimination of toxins as well as interference with normal healing and regenerative processes.

❖

RECOMMENDED REFERENCES

Becker, R.O., and Selden, G., **The Body Electric, Electromagnetism and the Foundation of Life,** (New York, NY: Morrow, 1985).Berkow, R. (ed), **The Merck Manual of Diagnosis and Therapy,** (Rahway, NJ: Merck, 1987).

Burr, H.S., **Blueprint for Immortality, The Electric Patterns of Life,** (Saffron Walden, Essex, England: The C.W. Daniel Co. Ltd., 1972).

Chaitow, L., and Kutter, E., **How to Live with Low-Level Radiation, A Nutritional Protection Plan,** (Rochester, VT: Healing Arts Press, 1988).

Coghill, R., **Electropollution, How to Protect Yourself Against It,** (Wellingborough, Northamptonshire, England: Thorson's Publishers Ltd., 1990).

Cohen, B., **Radon, A Homeowner's Guide to Detection and Control,** (Mount Vernon, NY: Consumer Reports Books, 1987).

Dakin, H.S., **High Voltage Photography,** (San Francisco, CA: H.S. Dakin Co., 1974).

Dinshah, D., **Let There Be Light,** (Malaga, NJ: Dinshah Health Society, 1985).

Fidler, J.H., **Earth Energy, A Dowser's Investigation of Ley Lines,** (Wellingborough, Northamptonshire, Great Britain: The Aquarian Press, 1988).

Garzino, S.J., **The Rhythm Factor in Human Behavior: The Challenges of Our Inner Rhythms, Clocks, and Cycles,** (Roslyn Heights, NY: Libra Publishers, Inc., 1982).

Gauquelin, M., **How Cosmic and Atmospheric Energies Influence Your Health,** (New York, NY: Aurora Press, 1971).

Gerber, R., **Vibrational Medicine, New Choices for Healing Ourselves,** (Santa Fe, NM: Bear & Company, 1988).

Gross, M.H., **Biorhythms,** (Albuquerque, NM: Motivation Development Center, 1975).

Habib, M.A., and Bockris, J. O., Interpretation of Current-Potential Relationships Across Biological Cell Membranes, in *Journal of Bioelectricity*, 1 (2) 289-294, (Marcel Dekker, Inc, 1982).

Hollwich, F., **The Influence of Ocular Light Perception on Metabolism in Man and in Animal,** (Heidelberg, Germany: Springer-Verlag, 1979).

Kenyon, J.N., **Modern Techniques of Acupuncture, Volume III, A Scientific Guide to Bio-electronic Regulatory Techniques and Complex Homeopathy,** (Wellingborough, Northamptonshire, Great Britain: Thorson's, 1985).

Kenyon, J.N., **21st Century Medicine, A Layman's Guide to the Medicine of the Future,** (Wellingborough, Northamtonshire, Great Britain: Thorson's, 1986).

Kern, K., **The Healthy House, An Owner-Builder's Guide to Biological Building,** (Oakhurst, CA: Owner-Builder Publications, 1978).

Kirschrink, J.L., Jones, D.S., and MacFadden, B.J. (eds), **Magnetite Biomineralization and Magnetoreception in Organisms. A New Biomagnetism,** (New York: Plenum, 1985).

Kønig, H.L., Bioinformation - Electrophysical Aspects, in (eds) Popp, F.A. and Becker, G., **Electromagnetic Bioinformation,** 25-54, (Munich: Urban and Schwartzenberg, 1979).

LeShan, L., **How to Meditate, A Guide to Self-Discovery,** (New York, NY: Bantam Books, 1974).

Liberman, J. I., **Light: Medicine of the Future,** (Santa Fe, NM: Bear & Co., 1991).

Luce, G.G., **Biological Rhythms in Human and Animal Physiology,** (New York: Dover Publications, 1971).

Lynes, B., **The Rife Report, The Cancer Cure That Worked, Fifty Years of Suppression,** (Queensville, Ontario, Canada: Marcus Books, 1987).

Mandel, P., **Energy Emission Analysis, New Application of Kirlian Photography for Holistic Health,** (Essen, Germany: Synthesis Publishing Company).

Mandel, P., **Practical Compendium of Colorpuncture,** (Bruchsal, Germany: Energetik Verlag, 1986).

Martina, R., **Futureplex Antitox Biotherapy, Physician's Desk Reference,** (Glendale, CA: Apex Energetics, 1988).

Martina, R., **Futureplex Causal Chain Therapy, Functional Medicine,** (Glendale, CA: Apex Energetics, 1989).

Miller, R.N., Methods of Detecting and Measuring Healing Energies, in **Future Science,** White and Krippner, Ed. (Anchor Books, 1977).

Ott, J.N., **Health and Light: The Effects of Natural and Artificial Light on Man and Other Living Things,** (New York, NY: Pocket Books, 1973).

Ott, J.N., **Light, Radiation, & You, How to Stay Healthy,** (Greenwich, CT: Devin-Adair Publishers, 1982).

Peat, F.D., **Superstrings and the Search for The Theory of Everything,** (Chicago: Contemporary, 1988).

Perry, S., and Dawson, J., **The Secrets Our Body Clocks Reveal,** (New York: Balantine Books, 1988).

Real Goods, pub., **Alternative Energy Sourcebook,** (Ukiah, CO: Real Goods, 1990).

Reckeweg, H.-H., **Homotoxicology, Illness and Healing through Anti-homotoxic Therapy,** (Baden-Baden, Germany: Menaco Publishing Company, Aurelia-Verlag, 1984).

Shannon, S., **Diet for the Atomic Age, How to Protect Yourself from Low-Level Radation,** (Wayne, NJ: Avery Publishing Group, Inc., 1987).

Sitchin, Z., **Genesis Revisited,** (New York, NY: Avon Books, 1990).

Smith, C.W., and Best, S., **Electromagnetic Man, Health and Hazard in the Electrical Environment,** (New York: St. Martin's,1989).

Soyka, F., **The Ion Effect, How Air Electricity Rules Your Life and Health,** (New York, NY: Bantam Books, 1977).

Spitler, H.R., **The Syntonic Principle,** (Lancaster, PA: Science Press,1941).

Stellman, J., and Henifin, M.S., **Office Work Can Be Dangerous to Your Health: A Handbook of Office Health and Safety Hazards and What You Can Do About Them,** (New York: Pantheon Books, 1983).

Stortebecker, P., **Dental Carries as a Cause of Nervous Disorders,** (Orlando, FL: Bio-Probe, Inc., 1986).

Stortebecker, P., **Mercury Poisoning from Dental Amalgam - a Hazard to Human Brain,** (Orlando, FL: Bio-Probe, Inc., 1986).

Swartwout, G., **Electromagnetic Health Research Manual,** (Hilo,HI: AERAI, 1990).

Webb, T., Lang, T., and Tucker, K., **Food Irradiation, Who Wants It?,** (Rochester, VT: Thorson's Publishers, Inc., 1987).

Wurtman, R.J., Baum, M.J., and Potts, J.T. (eds), **The Medical and Biological Effects of Light,** (New York, NY: The New York Academy of Sciences, 1985).

Zimmerman, R.L., Piezoelectricity and Biological Materials, in *J. Bioelectricity,* 1 (2), 265-287 (1982).

GLOSSARY

Achievement of Excellence Research Academy International (AERAI)

A membership research organization focusing on the investigation of stress elimination, including electromagnetic stresses, through exploration of natural means of self care.

AERAI (see Achievement of Excellence Research Academy International)

BEV (see Bioelectronics of Vincent)

Bioelectronics of Vincent (BEV)

BEV is an objective means of measuring and calculating biological energy based on both the **electrical** (electron) and **magnetic** (proton) factors in addition to the **ion** content or electrical conductivity (the inverse of resistivity). BEV measurement and analysis can be applied to blood, urine, saliva, water, nutrients, or other substances via measurement of the standard physical parameters (pH, rH_2 or ORP, and resistivity).

carrier frequency

The frequency or rate at which an oscillating pattern repeats itself. It acts as the carrier of the information contained in the **characteristic** pattern of the waveform. The carrier frequency determines the energy content of the individual photons which transmit the wave. A specific carrier frequency is like a specific radio station.

characteristic waveform

The characteristic waveform is the shape of an electromagnetic oscillation. It is determined by its specific source and represents an informational content of the electromagnetic oscillation. A characteristic waveform is like the programming on a radio station.

dowsing

Dowsing is an ancient art of finding water or other substances through amplification of subtle body responses. Dowsers may use wooden or metal dowsing rods, a **pendulum, radionic** instruments or other convenient amplification devices.

electromagnetic

Electromagnetic fields and radiations are composed of electrical and magnetic components. Electrostatic fields are produced by stationary electrical charges, such as the capacitor in a television set (even when unplugged). Magnetic fields are produced by electrical charges in constant motion, as in a direct current. When electrical charges change their pattern of motion, as in alternating current, electromagnetic radiation is produced.

electronic factor (rH$_2$)

The concentration of electrons in a fluid medium, such as in all biological systems is one of 3 factors which determine biological energy via the Nernst equation. The other two factors are the magnetic factor and the ion content (electrical conductivity). Because free electrons quickly combine with free protons to produce hydrogen (H$_2$), their concentration is measured as a function of hydrogen molecules (rH$_2$).

Hertz

The number of oscillations per second of an electromagnetic field is given the unit Hertz (Hz).

homeopathic

Homeopathically prepared substances are prepared by successive dilution and succussion to produce increasing potencies containing increasing electromagnetic carrier frequencies and decreasing chemical concentrations.

homotoxicology

Homotoxicology is the study of toxins in man. As toxins penetrate further into the system, they may enter more vital organs and tissues. They may also interfere at deeper levels within the cell and ultimately the nucleus. The reversal of this process is marked by a shift in symptoms to more superficial or less vital areas both anatomically and histologically. This detoxification process may also be marked by local metabolism of toxins accompanied inflammatory symptoms, and by increased elimination through mucus membranes, skin, urine or feces.

ion

A positively or negatively charge particle is an ion. Ions are capable of carrying electrical energy within the body by their movement. The total ion concentration determines the electrical conductivity of a fluid, which is one of 3 factors determining the total energy content according to the Nernst equation. The other two are the **electrical** and the **magnetic factors**.

magnetic factor (pH)

The concentration of protons, which are positively charged ions, is measured by the magnetic factor (pH). It is one of the 3 basic factors which determines the amount of biological energy via the Nernst equation. The other factors are the **electrical factor** and the **ion** content or electrical conductivity.

multi-dimensional

Multi-dimensional refers to any process which has more than just one or two dimensions or key factors. In truth, everything is multi-dimensional. It is only our limited perception, representation or thinking about something that can appear linear (1D) or flat (2D). Space is multi-dimensional (3D). Space-time, which Einstein showed to be inseparable except by the perception of each individual observer, is 4D. Physicists may now view the universe as 6D, 10D or 26D, depending on the context and model proposed.

pendulum

A pendulum is a simple device consisting of a suspended weight used to amplify subtle neuromuscular patterns in the arm for the detection of biological responses to subtle electromagnetic energy fields.

radionics

Radionics is an approach to detection of biological responses to subtle electromagnetic energy fields using a stick-plate which is rubbed by the finger tips. Amplification of subtle physiological changes takes place via noticable changes in the feel and sound of the stickiness of the stick-plate. Numbers called rates may be used to identify various fields, just as they might be identified by numbers such as frequencies, wavelenghts, or other characteristics of an electromagnetic oscillation.

Spectroscopy

A spectroscope measures the intensity of different frequencies of **electromagnetic** radiation. It is not able to measure the **characteristic** waveforms of the radiations. The information provided by spectroscopy is therefore like knowing what channels are on the air, but not what the programming is. The most sensitive indicators of the characteristic information in electromagnetic radiation is the physiological responses of biological systems.

Vega Biokinesiology (VBK)

VBK is an advanced method of muscle testing which integrates the protocols of the **Vegatest Method**.

Vegatest Method (VTM)

VTM is a method of electronic monitoring of skin resistance at an acupuncture point. **Homeopathic** stimuli are used to determine the patterns of causality and relief from stress at that time.

Vincent (see Bioelectronics of)

About the Author:

GLEN MARTIN SWARTWOUT

Glen was blessed with an upbringing that emphasized the natural approach to health and medicine taught by his family physician, Schuyler McCoulloch Martin, M.D., in honor of whom he received his middle name. Having worked for 20 years with Steinmetz and other top electromagnetic and bio-energetic researchers, Dr. Martin was truly a holistic clinician in his day and a pioneer in energy medicine. In the 1950's he taught Glen's mother, an R.N. and his father, an O.D. who himself helped to firmly establish the field of developmental optometry, about the dietary and lifestyle risk factors that only 30 years later began to enjoy broad acceptance in health care. Dr. Martin was clearly ahead of his time. Glen's service to others is offered with appreciation for the influences of Dr. Martin and his parents on his own access to biological energy, development and ability.

Achievement of Excellence
Research Academy International

MEMBERSHIP APPLICATION
(you may photocopy this form)

Some Benefits of Membership in AERAI :

1. Eligibility for confidential participation in state of the art research.

2. Opportunity for participation in personal support systems.

3. Placement on the Academy mailing list.

4. Receive a membership kit to get you started toward achieving excellence in body, mind and spirit.

5. Upon completion of your membership kit materials, you will receive several reports analyzing your present patterns of life stress and symptoms.

6. You will be entitled to a membership certificate, suitable for framing, upon successful completion of the membership kit materials.

7. Opportunities for learning, networking, personal growth and professional advancement.

Name: _____

Address: _____

_____

Phone: _____

Please enclose MEMBERSHIP FEE of $22, payable to:
AERAI, 100 Kam Ave #8, Hilo, Hawaii 96720.

Achievement of Excellence
Research Academy International

Glen Martin Swartwout,
A.B., B.D., O.D., F.I.C.A.N., F.C.S.O.

100 Kam Ave Suite 8 • Hilo HI 96720 • USA

Consulting 1-808-961-5593
Ordering 1-800-788-2442